THE BIRDS OF
JAPAN AND THE
BRITISH ISLES

A COMPARATIVE HAND LIST

OF

THE BIRDS OF
JAPAN AND THE BRITISH ISLES

BY

MASA U. HACHISUKA, F.Z.S.

MEMBER OF THE ORNITHOLOGICAL SOCIETY OF JAPAN,
MEMBER OF THE BRITISH ORNITHOLOGISTS' UNION

CAMBRIDGE
AT THE UNIVERSITY PRESS
1925

CAMBRIDGE
UNIVERSITY PRESS

University Printing House, Cambridge CB2 8BS, United Kingdom

Cambridge University Press is part of the University of Cambridge.

It furthers the University's mission by disseminating knowledge in the pursuit of
education, learning and research at the highest international levels of excellence.

www.cambridge.org
Information on this title: www.cambridge.org/9781107492905

© Cambridge University Press 1925

First published 1925
First paperback edition 2015

A catalogue record for this publication is available from the British Library

ISBN 978-1-107-49290-5 Paperback

TO

MY SISTER

VISCOUNTESS YASUHARU MATSUDAIRA

SOME NOTABLE WORKS ON THE ORNITHOLOGY OF JAPAN

TEMMINCK and SCHLEGEL. Siebold's "Fauna Japonica." *Aves*, 1844–1850.

BLAKISTON, T. "Ornithology of Northern Japan." *Ibis*, 1862, pp. 309–333.

BLAKISTON, T., and PRYER, H. "A Catalogue of the Birds of Japan." *Ibis*, 1878, pp. 209–250.

BLAKISTON, T., and PRYER, H. "A Catalogue of the Birds of Japan." *Trans. As. Soc. Japan*, 1880, pp. 172–241.

BLAKISTON, T., and PRYER, H. "Birds of Japan." *Trans. As. Soc. Japan*, 1882, pp. 84–186.

JOUY, P. L. "Notes on Birds from Japan." *Proc. U.S. Nat. Mus.* VI, pp. 273–318, 1883.

BLAKISTON, T. *Amended List of the Birds of Japan.* 1884.

STEJNEGER, L. "Review of Japanese Birds." *Proc. U.S. Nat. Mus.* IX, pp. 99–124, 1886.

,,	,,	,,	,,	,,	,,	,,	,,	IX, pp. 375–408, 1886.
,,	,,	,,	,,	,,	,,	,,	,,	X, pp. 271–319, 1887.
,,	,,	,,	,,	,,	,,	,,	,,	X, pp. 416–429, ,,
,,	,,	,,	,,	,,	,,	,,	,,	X, pp. 606–611, ,,
,,	,,	,,	,,	,,	,,	,,	,,	XI, pp. 425–432, 1888.
,,	,,	,,	,,	,,	,,	,,	,,	XI, pp. 547–548, ,,

SEEBOHM, H. *The Birds of the Japanese Empire.* 1890.

IJIMA, I. "Nippon no Torimokuroku (List of the Birds of Japan)." *Dobutsugaku-Zasshi (Tokyo Zool. Mag.)*, Vol. III, 1891.

STEJNEGER, L. "Notes on Birds from the Island of Yezo." *Proc. U.S. Nat. Mus.* XV, pp. 289–359, 1892.

INGRAM, C. "Ornithological Notes from Japan." *Ibis*, 1908, pp. 129–169.

OGAWA, M. "A Handlist of the Birds of Japan." *Annot. Zool. Japon.* Vol. VI, pp. 337–420, 1908.

CLARK, A. H. "Birds Collected and Observed in the North Pacific Ocean and in the Bering, Okhotsk, Japan and Eastern Seas." *Proc. U.S. Nat. Mus.* XXXVIII, pp. 25–74, 1910.

UCHIDA, S. "Nihon san Chorui Mokuroku (List of the Birds of Japan)." *Nihon Chorui Zusetsu (The Birds of Japan)*, Vol. II, pp. 1–36, 1914.

TAKA-TSUKASA, N., UCHIDA, S., KURODA, N., and MATSUDAIRA, Y. *A Hand-list of the Birds of Japan.* The Ornithological Society of Japan. 1922.

KURODA, N. "Notes on the Birds of Tsushima and Iki Islands." *Ibis*, 1922 pp. 75–105.

FOREWORD

IN this Hand List I have attempted to supply the materials for a complete comparison between the birds of the Island of Japan and the birds found in Great Britain and Ireland. These two island groups are situated in similar geographical positions, detached from the adjacent continents at the extreme ends of the Palæarctic Region. I have limited myself strictly to the Island of Japan—because the other parts of the Japanese Empire such as Korea, Tsushima, Formosa, Loo-choo Island and Bonin Island are zoo-geographically quite distinct. I have however included Tanegashima and Yakushima. In respect of the breeding records, the Japanese side has not as yet been completely worked out, but I have inserted all the instances which I believe to have been adequately attested. In classification I have chiefly followed Dr Hartert's *Die Vögel der Paläarktischen Fauna*, 1903– . I owe a great debt of gratitude to him, to Mr T. Momiyama and to Prince Taka-Tsukasa for their kind help in revising my manuscript. I have used three abbreviations throughout the work:

B.   denotes Breeding.
Exc.v.   „   Exceptional visitor.
R.v.   „   Rare visitor.

MASA U. HACHISUKA.

MAGDALENE COLLEGE,
    CAMBRIDGE.
October 1924.

CORRIGENDA

Order PASSERES

Family CORVIDÆ

1 Corvus Linnæus (1758)

JAPAN

1 *Corvus corax kamtschaticus* Dybowski.
Bull. Soc. Zool. France, 1883, pp. 362, 363.
Watari-garasu. (Kamschatkan Raven.) B.

2 *Corvus coronoides japonensis* Bonaparte.
Consp. Av. I, p. 386, 1850.
Hashibuto-garasu. (Japanese Jungle Crow.) B.

3 *Corvus corone interpositus* Laubmann.
Verh. Orn. Ges. Bayern, XIII, part 2, p. 201, 1917.
Hashiboso-garasu. (Japanese Carrion Crow.) B.

4 *Corvus frugilegus pastinator* (Gould).
P. Z. S., 1845, p. 1.
Miyama-garasu. (Eastern Rook.)

BRITISH ISLES

1 *Corvus corax corax* L.
Syst. Nat., ed. X, p. 105, 1758.
Raven. B.

2 *Corvus corone corone* L.
Syst. Nat., ed. X, p. 105, 1758.
Carrion Crow. B.

3 *Corvus cornix cornix* L.
Syst. Nat., ed. X, p. 105, 1758.
Hooded Crow. B.

4 *Corvus frugilegus frugilegus* L.
Syst. Nat., ed. X, p. 105, 1758.
Rook. B.

2 Colœus Kaup (1829)

5 *Colœus monedula spermologus* (Vieillot).
Nouv. Dict. d'Hist. Nat., VIII, p. 40, 1817.
Jackdaw. B.

5 *Colœus dauricus dauricus* (Pallas).
Reise Russ. R., III Anhang, p. 694, 1776.
Kokumaru-garasu. (Daurian Jackdaw.)

3 Pica Vieillot (1816)

6 *Pica pica sericea* Gould.
Proc. Zool. Soc. London, 1845, p. 2.
Kasasagi. (Chinese Magpie.) B.

7 *Pica pica bactriana* Bonaparte.
Consp. Av., I, p. 383, 1850.
Chishima-kasasagi. (White-rumped Himalayan Magpie.)

6 *Pica pica pica* (L.).
Syst. Nat., ed. X, p. 106, 1758.
Magpie. B.

4 **Cyanopica** Bonaparte (1850)

8 *Cyanopica cyanus japonica* Parrot.
Orn. Monatsber., 1905, p. 26.
  Onaga.   (Japanese   Azure-winged
  Magpie.) B.

5 **Garrulus** Vieillot (1816)

9 *Garrulus glandarius japonicus* Schle-
gel.
Fauna Japon., Aves, p. 83, pl. 43, 1848.
  Kakesu. (Japanese Jay.) B.

7 *Garrulus glandarius glandarius* (L.).
Syst. Nat., ed. X, p. 106, 1758.
  Continental Jay. (Exc.v.)

10 *Garrulus glandarius orii* Kuroda.
Bull. B. O. C., XLIII, p. 86, Jan. 1923.
  Yakushima-kakesu.   (Yakushima
  Jay.) B.

8 *Garrulus glandarius rufitergum* Har-
tert.
Vög. pal. Fauna, I, p. 30, 1903.
  British Jay. B.

11 *Garrulus glandarius brandtii* Evers-
mann.
Add. Pall. Zoogr., fasc. III, p. 8, 1842.
  Miyama-kakesu. (Brandt's Jay.) B.

9 *Garrulus glandarius hibernicus* Wither-
by and Hartert.
Brit. B., IV, p. 234, 1911.
  Irish Jay. B.

6 **Perisoreus** Bonaparte (1838)

12 *Perisoreus infaustus sakhalinensis* Bu-
turlin.
Mess. Ornith. Moscou, VII, p. 40, 1916.
  Akawo-kakesu. (Saghalien Island
  Jay.) B.

7 **Nucifraga** Vieillot (1816)

13 *Nucifraga caryocatactes japonicus* Har-
tert.
Nov. Zool., IV, p. 134, 1897.
  Hoshigarasu. (Japanese Nutcracker.)
  B.

10 *Nucifraga   caryocatactes   caryocatactes*
(L.).
Syst. Nat., ed. X, p. 106, 1758.
  Thick-billed Nutcracker. (R.v.)

14 *Nucifraga caryocatactes macrorhynchos*
C. L. Brehm.
Brehm, Lehrb. Naturg. europ., Vögel, I, p. 103,
1823.
  Hashinaga-hoshigarasu.   (Slender-
  billed Nutcracker.)

11 *Nucifraga caryocatactes macrorhynchos*
C. L. Brehm.
Brehm, Lehrb. Naturg. europ., Vögel, I, p. 103,
1823.
  Slender-billed Nutcracker.

8 **Pyrrhocorax** Vieillot (1816)

12 *Pyrrhocorax pyrrhocorax* (L.).
Syst. Nat., ed. X, p. 118, 1758.
  Chough. B.

Family STURNIDÆ

9 Sturnus Linnæus (1758)

13 *Sturnus vulgaris vulgaris* L.
Syst. Nat., ed. x, p. 167, 1758.
Starling. B.

14 *Sturnus vulgaris zetlandicus* Hartert.
Nov. Zool., 25, p. 329, 1918.
Shetland Starling. B.

10 Spodiopsar Sharpe (1889)

15 *Spodiopsar cineraceus* (Temminck).
Pl. Col., 11, pl. 556, 1835.
Mukudori. (Grey Starling.) B.

11 Pastor Temminck (1815)

15 *Pastor roseus* L.
Syst. Nat., ed. x, p. 170, 1758.
Rose-coloured Starling. (Exc.v.)

12 Sturnia Lesson (1837)

16 *Sturnia violacea* (Boddaert).
Tabl. Pl. Enl., p. 11, 1783.
Ko-mukudori. (Red-cheeked Myna.)
B.

17 *Sturnia sinensis* (Gmelin).
Syst. Nat., 1, p. 394, 1788.
Kara-mukudori.   (Chinese Myna.)
(Exc.v.)

Family ORIOLIDÆ

13 Oriolus Linnæus (1766)

16 *Oriolus oriolus* (L.).
Syst. Nat., ed. x, p. 107, 1758.
Golden Oriole. B.

Family FRINGILLIDÆ

14 Chloris Cuvier (1800)

17 *Chloris chloris chloris* (L.).
Syst. Nat., ed. x, p. 174, 1758.
Greenfinch. B.

18 *Chloris sinica kawarahiba* (Temminck).
Pl. Color., 3, pl. 588, fig. 1, 1836.
O-kawarahiwa. (Japanese Large
Greenfinch.) B.

19 *Chloris sinica ussuriensis* Hartert.
Vög. pal. Fauna, I, p. 64, 1903.
    Chosen-kawarahiwa. (Ussur Green-
finch.) B.

20 *Chloris sinica sitchitoensis* Momiyama.
Tokyo Zool. Mag., Feb. 1924.
    Shichito-kawarakiwa. (Seven Island
Greenfinch.) B.

21 *Chloris sinica minor* (Temminck and
Schlegel).
Fauna Japon., Aves, pl. 49, p. 89, 1850.
    Ko-kawarahiwa. (Japanese Lesser
Greenfinch.) B.

15 Coccothraustes Pallas (1811)

22 *Coccothraustes coccothraustes japonicus*
(Temminck and Schegel).
Fauna Japon., Aves, pl. 51, 1850.
    Shime. (Japanese Hawfinch.) B.

18 *Coccothraustes coccothraustes coccothraus-*
tes (L.).
Syst. Nat., ed. X, p. 171, 1758.
    Hawfinch. B.

23 *Coccothraustes coccothraustes verticalis*
Tugarinow and Buturlin.
Mater, Vög. Jeness. Gouv., in Mitt. Krasno-
jarsker Sektion Russ. Geogr. Ges., 1911.
    Karafuto-shime. (Eastern Hawfinch.)

16 Eophona Gould (1851)

24 *Eophona personatus personatus* (Temminck and Schlegel).
Fauna Japon., Aves, p. 91, pl. 52, 1850.
    Ikaru. (Japanese Hawfinch.) B.

25 *Eophona melanura migratoria* Har-
tert.
Vög. pal. Fauna, I, p. 59, fig. 17, 1903.
    Ko-ikaru. (Migratory Black-tailed
Hawfinch.)

17 Carduelis Brisson (1760)

19 *Carduelis carduelis carduelis* (L.).
Syst. Nat., ed. X, p. 180, 1758.
    Continental Goldfinch.

20 *Carduelis carduelis britannicus* (Hartert).
Vög. pal. Fauna, I, p. 68, 1903.
    British Goldfinch. B.

18 Spinus Koch (1816)

26 *Spinus spinus* (L.).
Syst. Nat., ed. X, p. 181, 1758.
Mahiwa. (Siskin.) B.

21 *Spinus spinus* (L.).
Syst. Nat., ed. X, p. 181, 1758.
Siskin.

22 *Spinus citrinella citrinella* (L.).
Syst. Nat., ed. XII, p. 320, 1766.
Citril Finch. (Exc.v.)

19 Serinus Koch (1816)

23 *Serinus canarius serinus* (L.).
Syst. Nat., ed. XII, p. 320, 1766.
Serin. (R.v.)

20 Passer Koch (1816)

24 *Passer domesticus domesticus* (L.).
Syst. Nat., ed. X, p. 183, 1758.
House-Sparrow. B.

27 *Passer montanus kaibatoi* Munster-hjelm.
Nyt. Mag. for Naturvidensk. 1916, p. 170.
Karafuto-suzume. (Saghalien Island Tree-Sparrow.) B.

25 *Passer montanus montanus* (L.).
Syst. Nat., ed. X, p. 183, 1758.
Tree-Sparrow. B.

28 *Passer montanus saturatus* Stejneger.
Proc. U.S. Nat. Mus., VIII, p. 19, 1885.
Suzume. (Japanese Tree-Sparrow.) B.

29 *Passer rutilans rutilans* (Temminck).
Pl. Color., III, 488, 1829.
Niunai-suzume. (Russet Sparrow.) B.

21 Montifringilla Brehm (1828)

26 *Montifringilla nivalis nivalis* (L.).
Syst. Nat., ed. XII, p. 321, 1766.
Snow-Finch. (R.v.)

30 *Montifringilla brunneinucha* (Brandt).
Bull. Sci. Acad. St Petersb., X, p. 252, 1842.
Hagi-mashiko. (Japanese Ground-Linnet.) B.

22 Fringilla Linnæus (1758)

27 *Fringilla cœlebs cœlebs* L.
Syst. Nat., ed. X, p. 179, 1758.
Chaffinch. B.

31 *Fringilla montifringilla* L.
Syst. Nat., ed. X, p. 179, 1758.
Atori. (Brambling.)

28 *Fringilla montifringilla* L.
Syst. Nat., ed. X, p. 179, 1758.
Brambling. B.

23 Acanthis Bechstein (1803)

29 *Acanthis cannabina cannabina* (L.).
Syst. Nat., ed. x, p. 182, 1758.
Linnet. B.

30 *Acanthis linaria linaria* (L.).
Syst. Nat., ed. x, p. 182, 1758.
Mealy Redpoll.

31 *Acanthis linaria rostrata* (Coues).
Proc. Ac. Nat. Sci. Philadelphia, Nov., 1861
(publ. 1862), p. 378.
Greenland Redpoll. (Exc.v.)

32 *Acanthis linaria holboelli* (C. L. Brehm).
Handb. Naturg. Vög. Deutschl., p. 280, 1831.
Beni-hiwa. (Holböll's Redpoll.)

32 *Acanthis linaria holboellii* (C. L. Brehm).
Handb. Naturg. Vög. Deutschl., p. 280, 1831.
Holböll's Redpoll. (Exc.v.)

33 *Acanthis linaria cabaret* (P. L. S. Müller).
Natursystem, Suppl., p. 165, 1776.
Lesser Redpoll. B.

34 *Acanthis hornemanni hornemanni* (Holböll).
Naturh. Tidskr., iv, p. 398, 1843.
Hornemann's Redpoll. (R.v.)

33 *Acanthis hornemanni exilipes* (Coues).
Proc. Ac. Nat. Sci. Philadelphia, Nov., 1861,
(pub. 1862), p. 385.
Ko-beni-hiwa. (Coues's Redpoll.)

35 *Acanthis hornemanni exilipes* (Coues).
Proc. Ac. Nat. Sci. Philadelphia, Nov., 1861
(pub. 1862), p. 385.
Coues's Redpoll. (Exc.v.)

36 *Acanthis flavirostris flavirostris* (L.).
Syst. Nat., ed. x, p. 182, 1758.
Twite. B.

24 Pyrrhula Pallas (1811)

34 *Pyrrhula pyrrhula kamtschatica* Taczanowski.
Bull. Soc. Zool. France, 1882, p. 395.
Benibara-uso. (Kamschatkan Bullfinch.) (R.v.)

37 *Pyrrhula pyrrhula pyrrhula* (L.).
Syst. Nat., ed. x, p. 171, 1758.
Northern Bullfinch. (V.)

35 *Pyrrhula pyrrhula kurilensis* Sharpe.
Zoologist, 1886, p. 485.
Chishima-uso. (Kurile Island Bullfinch.) B.

38 *Pyrrhula pyrrhula nesa* Mathews and Iredale.
Austr. Av. Record, iii, p. 122, 1917.
British Bullfinch. B.

36 *Pyrrhula pyrrhula griseiventris* Lafresnaye.
Rev. Zool., 1841, p. 241.
Uso. (Oriental Bullfinch.) B.

25 Uragus Keys and Blasius (1840)

37 *Uragus sibirica sanguinolenta* (Temminck and Schlegel).
Fauna Japon., Aves, p. 92, pl. 54, 1850.
Beni-mashiko. (Japanese Rose Finch.) B.

26 Carpodacus Kaup (1829)

38 *Carpodacus rosea* (Pallas).
Reise d. versch. Prov. d. Russ. Reichs, III, p. 699, 1776.
O-mashiko. (Pallas's Rose Finch.)

39 *Carpodacus erythrinus grebnitskii* Stejneger.
Orn. Expl. Command Is. and Kamschatka, p. 265, 1885.
Aka-mashiko. (Kamschatkan Scarlet Finch.)

39 *Carpodacus erythrinus erythrinus* (Pallas).
Nov. Comm. Acad. Sci. Imp. Petropol., XIV, p. 587, pl. 23, fig. 1, 1770.
Scarlet Grosbeak. (R.v.)

27 Pinicola Vieillot (1807)

40 *Pinicola enucleator sakhalinensis* Buturlin.
Mess. Orn., 1915, pp. 129, 130.
Karafuto-ginzan-mashiko. (Saghalien Island Pine-Grosbeak.) B.

40 *Pinicola enucleator enucleator* (L.).
Syst. Nat., ed. x, p. 171, 1758.
Pine Grosbeak. (R.v.)

41 *Pinicola enucleator urupensis* Buturlin.
Mess. Orn., 1915, p. 239.
Ginzan-mashiko. (Kurile Island Pine-Grosbeak.) B.

28 Loxia Linnæus (1758)

42 *Loxia curvirostra caucasica* Buturlin.
Orn. Monatsber., 1907, p. 9.
Isuka. (Eastern Crossbill.)

41 *Loxia curvirostra curvirostra* L.
Syst. Nat., ed. x, p. 171, 1758.
Common Crossbill. B.

43 *Loxia curvirostra japonica* Ridgway.
Proc. Biol. Soc. Washington, II, p. 101, 1885.
Shirohara-isuka. (Japanese Crossbill.) B.

42 *Loxia curvirostra scotica* Hartert.
Vög. pal. Fauna, I, p. 120, fig. 24, 1904.
Scottish Crossbill. B.

43 *Loxia pytyopsittacus* Borkhausen.
Rheinisches Magazin, I, p. 139, 1793.
Parrot Crossbill. (R.v.)

44 *Loxia leucoptera elegans* Homeyer.
Journ. f. Orn., 1879, p. 180.
Naki-isuka. (Eastern Two-barred Crossbill.)

44 *Loxia leucoptera bifasciata* (C. L. Brehm).
Ornis, III, p. 85, 1827.
Two-barred Crossbill. (R.v.)

45 *Loxia leucoptera leucoptera* Gmelin.
Syst. Nat., I, 2, p. 844, 1789.
American White-winged Crossbill. (Exc.v.)

29 **Emberiza** Linnæus (1758)

46 *Emberiza calandra calandra* L.
Syst. Nat., ed. **x**, p. 176, 1758.
Corn-Bunting. B.

47 *Emberiza citrinella citrinella* L.
Syst. Nat., ed. **x**, p. 177, 1758.
Yellow Bunting. B.

45 *Emberiza leucocephala* S. G. Gmelin.
Nov. Comm. Acad. Sci. Imp. Petropol., **xv**,
p. 480, pl. 23, fig. 3, 1771.
Shiraga-hōjiro. (Pine-Bunting.) B.

48 *Emberiza leucocephala* (S. G. Gmelin).
Nov. Comm. Acad. Sci. Imp. Petropol., **xv**,
p. 480, pl. 23, fig. 3, 1771.
Pine-Bunting. (Exc.v.)

49 *Emberiza cirlus* L.
Syst. Nat., ed. **xii**, p. 311, 1766.
Cirl Bunting. B.

50 *Emberiza melanocephala* Scopoli.
Annus I, Hist. Nat., p. 142, 1769.
Black-headed Bunting. (Exc.v.)

51 *Emberiza cia cia* L.
Syst. Nat., ed. **xii**, p. 310, 1766.
Meadow or Rock-Bunting. (Exc.v.)

46 *Emberiza rutila* Pallas.
Reise d. versch. Prov. d. Russ. Reichs, **iii**,
p. 698, 1776.
Shima-nojiko. (Chestnut Bunting.)
(R.v.)

47 *Emberiza aureola* Pallas.
Reise d. versch. Prov. d. Russ. Reichs, **ii**
Anhang, p. 711, 1773.
Shima-awoji. (Yellow-breasted Bunt-ing.) B.

52 *Emberiza aureola* (Pallas).
Reise d. versch. Prov. d. Russ. Reichs, **ii**
Anhang, p. 711, 1773.
Yellow-breasted Bunting. (Exc.v.)

48 *Emberiza elegans* Temminck.
Pl. Color., 583, 1835.
Miyama-hojiro. (Yellow-throated
Bunting.) B.

49 *Emberiza spodocephala spodocephala*
Pallas.
Reise d. versch. Prov. d. Russ. Reichs, **iii**,
p. 698, 1776.
Kara-awoji. (Black-faced Bunting.)
(Exc.v.)

50 *Emberiza spodocephala personata* Tem-minck.
Pl. Color., 580, 1835.
Awoji. (Japanese Bunting.) B.

51 *Emberiza sulphurata* Temminck and Schlegel.
Fauna Japonica, Aves, p. 100, pl. 60, 1848.
Nojiko. (Japanese Yellow Bunting.) B.

52 *Emberiza cioides ciopsis* Bonaparte.
Consp. Av., I, p. 466, 1850.
Hojiro. (Japanese Meadow-Bunting.) B.

53 *Emberiza cioides namiyei* Momiyama.
Tori, Vol. III, No. 4, p. 210, 1923.
Namiye-hojiro. (Namiye's Meadow-Bunting.) B.

54 *Emberiza cioides tametomo* Momiyama.
Tokyo Zool. Mag., Feb. 1924.
Tametomo-hojiro. (Momiyama's Meadow-Bunting.) B.

55 *Emberiza cioides ijimae* Stejneger.
Proc. U.S. Nat. Mus., XVI, p. 638, 1894.
Ijima-hojiro. (Stejneger's Meadow-Bunting.) B.

56 *Emberiza cioides neglecta* Kuroda.
Bull. B.O.C., XLIII, p. 88, Jan. 1923.
Yakushima-hojiro. (Kuroda's Meadow-Bunting.) B.

53 *Emberiza cioides castaneiceps* Moore.
Proc. Zool. Soc. London, 1855, p. 215.
East Siberian Meadow-Bunting. (Exc.v.)

54 *Emberiza hortulana* L.
Syst. Nat., ed. x, p. 177, 1758.
Ortolan Bunting. (R.v.)

57 *Emberiza fucata fucata* Pallas.
Reise d. versch. Prov. d. Russ. Reichs, III, p. 698, 1766.
Hōaka. (Grey-headed Bunting.) B.

58 *Emberiza pusilla* Pallas.
Reise d. versch. Prov. d. Russ. Reichs, III, p. 697, 1776.
Ko-hōaka. (Little Bunting.) (R.v.)

59 *Emberiza rustica* Pallas.
Reise d. versch. Prov. d. Russ. Reichs, III, p. 698, 1776.
Kashiradaka. (Rustic Bunting.)

60 *Emberiza schœniclus pyrrhulinus* Swinhoe.
Ibis, 1876, p. 333, pl. 8, fig. 2.
O-jurin. (Finch-billed Reed-Bunting.) B.

55 *Emberiza pusilla* Pallas.
Reise d. versch. Prov. d. Russ. Reichs, III, p. 697, 1776.
Little Bunting. (R.v.)

56 *Emberiza rustica* Pallas.
Reise d. versch. Prov. d. Russ. Reichs, III, p. 698, 1776.
Rustic Bunting. (R.v.)

57 *Emberiza schœniclus schœniclus* (L.).
Syst. Nat., ed. x, p. 182, 1758.
Reed-Bunting. B.

58 *Emberiza tschusii tschusii* Reiser and Almasy.
   Aquila, v, pp. 122–5, 1898.
      Eastern Large-billed Reed-Bunting. (R.v.)

59 *Emberiza tschusii compilator* Mathews and Iredale.
   Austr. Av. Record, IV, p. 131, 1920.
      Western Large-billed Reed-Bunting. (R.v.)

61 *Emberiza variabilis* Temminck.
   Pl. Color., 583, 1835.
      Kuroji. (Japanese Grey Bunting.) B.

62 *Emberiza yessoënsis yessoënsis* (Swin-hoe).
   Ibis, 1874, p. 161 (ex Blakiston, Ibis, 1863, p. 99).
      Ko-jurin. (Swinhoe's Japanese Bunt-ing.) B.

30 **Calcarius** Bechstein (1803)

63 *Calcarius lapponicus coloratus* Ridg-way.
   Auk, xv, p. 320, 1898.
      Tsumenaga-hōjiro.   (Kamschatkan Bunting.)

60 *Calcarius lapponicus lapponicus* (L.).
   Syst. Nat., ed. x, p. 180, 1758.
      Lapland Bunting.

31 **Plectrophenax** Stejneger (1882)

64 *Plectrophenax nivalis townsendi* Ridg-way.
   Manual of N. Amer. Birds, p. 403, 1887.
      Yuki-hōjiro. (Prybilof Snow-Bunt-ing.) (R.v.)

61 *Plectrophenax nivalis nivalis* (L.).
   Syst. Nat., ed. x, p. 176, 1758.
      Snow-Bunting. B.

Family ALAUDIDÆ

32 **Alauda** Linnæus (1758)

65 *Alauda arvensis pekinensis* Swinhoe.
   Proc. Zool. Soc. London, 1863, p. 89.
      O-hibari. (Blakiston's Sky-Lark.) B.

62 *Alauda arvensis arvensis* L.
   Syst. Nat., ed. x, p. 165, 1758.
      Sky-Lark. B.

66 *Alauda arvensis intermedia* Swinhoe.
   Proc. Zool. Soc. London, 1863, p. 89.
      Chiu-hibari. (Eastern Sky-Lark.) B.

63 *Alauda arvensis intermedia* Swinhoe.
   Proc. Zool. Soc. London, 1863, p. 89.
      Eastern Sky-Lark. (R.v.)

67 *Alauda arvensis japonica* Temminck and Schlegel.
   In Siebold's Fauna Japonica, Aves, p. 87, pl. 47, 1848.
      Hibari. (Japanese Sky-Lark.) B.

33 Lullula Kaup (1829)

64 *Lullula arborea arborea* (L.).
Syst. Nat., ed. x, p. 166, 1758.
Wood-Lark. B.

34 Galerida Boie (1828)

65 *Galerida cristata cristata* (L.).
Syst. Nat., ed. x, p. 166, 1758.
Crested-Lark. (R.v.)

35 Calandrella Kaup (1829)

66 *Calandrella brachydactyla brachydactyla*
(Leisler).
Ann. Wetterau. Ges., III, p. 357, pl. 19, 1814.
Short-toed Lark. (R.v.)

67 *Calandrella brachydactyla longipennis*
(Eversmann).
Bull. Soc. Imp. Nat. Moscou, XXI, p. 219, 1848.
Eastern Short-toed Lark. (R.v.)

36 Melanocorypha Boie (1828)

?68 *Melanocorypha bimaculata* (Ménétries).
Cat. Rais., p. 37, 1832.
Kubiwa-kotenshi. (Eastern Calandra
Lark.) (Exc.v.)

68 *Melanocorypha sibirica* (Gmelin).
Syst. Nat., I, 2, p. 799, 1789.
White-winged Lark. (R.v.)

69 *Melanocorypha calandra calandra* (L.).
Syst. Nat., ed. XII, p. 288, 1766.
Calandra Lark. (R.v.)

70 *Melanocorypha yeltoniensis* (Forster).
Philos. Trans., LVII, p. 350, 1768.
Black Lark.

37 Eremophila Boie (1828)

69 *Eremophila alpestris enroa* (Thayer and
Bangs).
Proc. New England Zool. Club, V, p. 43, 1914.
Hama-hibari. (East Siberian Shore-
Lark.) (R.v.)

71 *Eremophila alpestris flava* (Gmelin).
Syst. Nat., I, 2, p. 800, 1789.
Shore-Lark.

Family MOTACILLIDÆ

38 Motacilla Linnæus (1758)

72 *Motacilla alba alba* L.
Syst. Nat., ed. x, p. 185, 1758.
White Wagtail. B.

70 *Motacilla alba lugens* Kittlitz.
Kupfertafeln zur Naturg. d. Vög., part 2, p. 16,
pl. 21, fig. 1, 1833.
Haku-sekirei. (Japanese Pied Wagtail.) B.

73 *Motacilla alba yarrellii* Gould.
Birds of Europe, List of Plates, Vol. II, p. 2,
1837.
Pied Wagtail. B.

71 *Motacilla alba grandis* Sharpe.
Cat. B. Brit. Mus., x, p. 492, 1885.
Seguro-sekirei. (Japanese Wagtail.)
B.

72 *Motacilla cinerea melanope* Pallas.
Reise d. versch. Prov. d. Russ. Reichs, III,
p. 696, 1776.
Kisekirei. (Eastern Grey Wagtail.)
B.

73 *Motacilla flava simillima* Hartert.
Vog. pal. Fauna, vol. I, p. 289, 1905.
Mamijiro-tsumenaga-sekirei. (East-
ern Blue-headed Wagtail.) (R.v.)

74 *Motacilla flava taivanus* (Swinhoe).
Proc. Zool. Soc. London, 1863, p. 334.
Tsumenaga-sekirei. (Eastern Yellow
Wagtail.) B.

74 *Motacilla alba personata* Gould.
B. Asia, IV, pl. 63, 1861.
Masked Wagtail. (R.v.)

75 *Motacilla cinerea cinerea* Tunstall.
Orn. Brit., p. 2, 1771.
Grey Wagtail. B.

76 *Motacilla flava flava* L.
Syst. Nat., ed. x, p. 185, 1758.
Blue-headed Wagtail. B.

77 *Motacilla flava rayi* (Bonaparte).
Geog. and Comp. List of B. Europe and
N. America, p. 18, 1838.
Yellow Wagtail. B.

78 *Motacilla flava beema* (Sykes).
Proc. Committee Zool. Soc., London, part II,
1832, p. 90.
Sykes's Wagtail (Exc.v.)

79 *Motacilla flava cinereocapilla* Savi.
Nuovo Giornale dei Letterati, no. 57, p. 190,
1831.
Ashy-headed Wagtail. (Exc.v.)

80 *Motacilla flava thunbergi* Billberg.
Billberg, Synopsis Fauna Scand., I, 2, Aves,
p. 50, 1828.
Grey-headed Wagtail.

81 *Motacilla flava feldegg* Michahelles.
Isis, 1830, p. 812.
Black-headed Wagtail. (Exc.v.)

39 Dendronanthus Blyth (1844)

75 *Dendronanthus indicus* (Gmelin).
Syst. Nat., I, 2, p. 962, 1789.
Iwami-sekirei. (Indian Wagtail.)
(Exc.v.)

40 Anthus Bechstein (1807)

76 *Anthus trivialis hodgsoni* Richmond.
Carnegie Insti. Washington, No. 54, p. 493,
1907.
Binzui. (Eastern Tree-Pipit.) B.

82 *Anthus trivialis trivialis* (L.).
Syst. Nat., ed. x, p. 166, 1758.
Tree-Pipit. B.

83 *Anthus pratensis* (L.).
Syst. Nat., ed. x, p. 166, 1758.
Meadow-Pipit. B.

77 *Anthus cervinus* (Pallas).
Zoogr. Rosso-Asiat., I, p. 511, 1827.
Muneaka-tahibari. (Red-throated
Pipit.) B.

84 *Anthus cervinus* (Pallas).
Zoogr. Rosso-Asiat., I, p. 511, 1827.
Red-throated Pipit. (R.v.)

85 *Anthus campestris* (L.).
Syst. Nat., ed. x, p. 166, 1758.
Tawny Pipit. B.

86 *Anthus richardi richardi* Vieillot.
Nouv. Dict. d'Hist. Nat., nouv. xxvi, p. 491,
1818.
Richard's Pipit.

78 *Anthus spinoletta borealis* Hesse.
Journ. f. Orn., 1915, p. 386.
Karafuto-tahibari. (Saghalien Island
Water-Pipit.)

87 *Anthus spinoletta spinoletta* (L.).
Syst. Nat., ed. x, p. 166, 1758.
Water-Pipit.

79 *Anthus spinoletta japonicus* Temminck
and Schlegel.
Siebold's Fauna Japon., Aves, p. 59, pl. 24,
1847.
Tahibari. (Japanese Water-Pipit.) B.

88 *Anthus spinoletta rubescens* (Tunstall).
Orn. Brit., p. 2, 1771.
American Water-Pipit. (Exc.v.)

89 *Anthus spinoletta petrosus* (Montagu).
Trans. Linn. Soc. London, IV, p. 41, 1798.
Rock-Pipit. B.

90 *Anthus spinoletta littoralis* Brehm.
Handb. Naturg. Vög. Deutschl., p. 331, 1831.
Scandinavian Rock-Pipit.

Family MUSCICAPIDÆ

41 **Tchitrea** Lesson (1831)

80 *Tchitrea atrocaudata atrocaudata* (Ey-
ton).
Proc. Zool. Soc. London, 1839, p. 102.
Sankocho. (Japanese Paradise Flycatcher.) B.

42 **Muscicapa** Linnæus 1766

91 *Muscicapa striata striata* (Pallas).
Vroeg's Cat. Verzam. Vogelen, etc., Adum-
bratiuncula, p. 3, 1764.
Spotted Flycatcher. B.

81 *Muscicapa latirostris* Raffles.
Trans. Linn. Soc. London, XIII, 2, p. 312, 1821.
Ko-samebitaki. (Brown Flycatcher.)
B.

92 *Muscicapa latirostris* Raffles.
Trans. Linn. Soc. London, XIII, 2, p. 312, 1821.
Brown Flycatcher. (Exc.v.)

43 Ficedula Brisson (1760)

93 *Ficedula hypoleuca hypoleuca* (Pallas).
Vroeg's Cat. Verzam. Vogelen, etc., Adumbratiuncula, p. 3, 1764.
Pied Flycatcher. B.

94 *Ficedula albicollis* (Temminck).
Man. d'Orn., p. 100, 1815.
Collared Flycatcher. (Exc.v.)

44 Siphia Hodgson (1837)

95 *Siphia parva parva* (Bechstein).
Latham's Allg. Uebers. d. Vögel, II, I, p. 356, fig. on title-page, 1794.
Red-breasted Flycatcher.

45 Hemichelidon Hodgson (1845)

82 *Hemichelidon sibirica sibirica* Gmelin.
Syst. Nat., I, 2, p. 936, 1789.
Same-bitaki. (Siberian Flycatcher.)
B.

83 *Hemichelidon griseisticta griseisticta* Swinhoe.
Ibis, 1861, p. 330.
Yezo-bitaki. (Swinhoe's Flycatcher.)

84 *Hemichelidon griseisticta habereri* (Parrot).
Ornith. Monatsber. Berlin, xv, pp. 168–170, 1907.
Chishima-bitaki. (Kurile Island Flycatcher.)

46 Zanthopygia Blyth (1847)

85 *Zanthopygia narcissina narcissina* (Temminck).
Pl. Color., 577, fig. 1, 1835.
Kibitaki. (Narcissus Flycatcher.) B.

86 *Zanthopygia narcissina jakuschima* (Hartert).
Vog. pal. Fauna, I, p. 491, 1907.
Yakushima-kibitaki. (Yakushima Narcissus Flycatcher.) B.

87 *Zanthopygia narcissina zanthopygia* (Hay).
Madras Journal, XIII, 2, p. 162, 1845.
Mamijiro-kibitaki. (Hay's Narcissus Flycatcher.) (Exc.v.)

47 Poliomyias Sharpe (1897)

88 *Poliomyias mugimaki* (Temminck).
Pl. Color., 577, fig. 2, 1835.
Mugimaki. (Mugimaki Flycatcher.)
B.

48 Cyanoptila Blyth (1847)

89 *Cyanoptila cyanomelana* (Temminck).
Pl. Color., 470, 1828.
Oruri. (Japanese Blue Flycatcher.) B.

Family BRACHYPODIDÆ

49 Microscelis Gray (1840)

90 *Microscelis amaurotis amaurotis* (Temminck).
Pl. Color., II, pl. 497, 1830.
Hiyodori. (Brown-eared Bulbul.) B.

91 *Microscelis amaurotis hensoni* (Stejneger).
Proc. U.S. Nat. Mus., xv, p. 347, 1893.
Ezo-hiyodori. (Stejneger's Bulbul.) B.

92 *Microscelis amaurotis matchie* Momiyama.
Tokyo Zool. Mag., Feb. 1924.
Hachijo-hiyodori. (Seven Island Bulbul.) B.

Family ZOSTEROPIDÆ

50 Zosterops Vigors and Horsfield (1826)

93 *Zosterops palpebrosa japonicus* Temminck and Schlegel.
In Siebold's Fauna Japonica, Aves, p. 57, pl. 22, 1848.
Mejiro. (Japanese White-Eye.) B.

94 *Zosterops palpebrosa stejnegeri* Seebohm.
Ibis, 1891, p. 273.
Shichito-mejiro. (Seven Island White-Eye.) B.

95 *Zosterops palpebrosa ijimae* Kuroda.
Tori, No. 5, p. 4, pl. vi, fig. 3, text fig. 2, 1917.
Ijima-mejiro. (Ijima's White-Eye.) B.

96 *Zosterops palpebrosa insularis* Ogawa.
Annot. Zool. Japon., v, p. 186, 1905.
Shima-mejiro. (Ogawa's White-Eye.) B.

Family CERTHIDÆ

51 Certhia Linnæus (1758)

97 *Certhia familiaris familiaris* L.
Syst. Nat., ed. X, p. 118, 1758.
Kita-kibashiri. (Northern Tree-Creeper.) B.

98 *Certhia familiaris japonica* Hartert.
Nov. Zool., 1897, p. 138.
Kibashiri. (Japanese Tree-Creeper.) B.

96 *Certhia familiaris familiaris* L.
Syst. Nat., ed. X, p. 118, 1758.
Northern Tree-Creeper. (R.v.)

97 *Certhia familiaris brittanica* Ridgway.
Proc. U.S. Nat. Mus., V, p. 113, 1882.
British Tree-Creeper. B.

52 Tichodroma Illiger (1811)

98 *Tichodroma muraria* (L.).
Syst. Nat., ed. XII, p. 184, 1766.
Wall-Creeper. (R.v.)

Family SITTIDÆ

53 Sitta Linnæus (1758)

99 *Sitta europæa albifrons* Taczanowski.
Bull. Soc. Zool. France, VII, p. 385, 1882.
Shirobitae-gojukara. (Kamschatkan Nuthatch.)

100 *Sitta europæa sakhalinensis* Buturlin.
Trav. Soc. Nat. Petrograd, XLIV, p. 170, 1916.
Shirohara-gojukara. (Saghalien Island Nuthatch.) B.

101 *Sitta europæa clara* Stejneger.
Proc. U.S. Nat. Mus., IX, pp. 390, 392, 1886.
Yezo-gojukara. (Yesso Nuthatch.) B.

102 *Sitta europæa hondoensis* Buturlin.
Trav. Soc. Nat. Petrograd, XLIV, p. 171, 1916.
Gojukara. (Hondo Nuthatch.) B.

99 *Sitta europæa affinis* Blyth.
Journ. Asiatic Soc. Bengal, XV, p. 289, 1846.
British Nuthatch. B.

Family PARIDÆ

54 Regulus Vieillot (1807)

103 *Regulus regulus japonensis* Blakiston.
Ibis, 1862, p. 320.
Kikuitadaki. (Eastern Golden-crested Wren.) B.

100 *Regulus regulus regulus* (L.).
Syst. Nat., ed. X, p. 188, 1758.
Continental Golden-crested Wren.

101 *Regulus regulus anglorum* Hartert.
Bull. B. O. C., XVI, p. 11, Oct. 1905.
British Golden-crested Wren. B.

102 *Regulus ignicapillus ignicapillus* (Temminck).
Man. d'Orn., ed. II, I, p. 231, 1820.
Fire-crested Wren.

55 Panurus Koch (1816)

104 *Panurus biarmicus russicus* (Brehm).
Handb. Naturg. Vög. Deutschl., p. 472, 1831.
Higegara. (Eastern Bearded Titmouse.) (R.v.)

103 *Panurus biarmicus biarmicus* (L.).
Syst. Nat., ed. X, p. 190, 1758.
Bearded Titmouse. B.

56 Parus Linnæus (1758)

105 *Parus major minor* Temminck and Schlegel.
In Siebold's Fauna Japonica, Aves, p. 70, pl. 33, 1848.
Shijukara. (Japanese Titmouse.) B.

104 *Parus major major* L.
Syst. Nat., ed. X, p. 189, 1758.
Continental Great Titmouse.

106 *Parus major ogawai* Momiyama.
Tori, Vol. III, No. 14, p. 207, 1923.
Ogawa-shijukara. (Ogawa's Titmouse.) B.

105 *Parus major newtoni* Prazak.
Orn. Jahrb., v, p. 239, 1894.
British Great Titmouse. B.

107 *Parus major quelpartensis* Kuroda.
Tori, I, No. 5, p. 3, pl. VI, figs. 1, 2, 1917.
Shima-shijukara. (Quelpart Island Titmouse.) B.

108 *Parus major chimae* Momiyama.
Tokyo Zool. Mag., Feb. 1924.
Hachijo-shijukara. (Hachijo Island Titmouse.) B.

109 *Parus major kagoshimae* Takatsukasa.
Dobutsugaku Zasshi, XXXI, p. 55, 1919.
Kagoshima-shijukara. (Takatsukasa's Titmouse.) B.

110 *Parus ater amurensis* (Buturlin).
Orn. Monatsb., 1907, p. 80.
Kita-higara. (Amur Coal-Titmouse.) B.

106 *Parus ater ater* L.
Syst. Nat., ed. X, p. 190, 1758.
Continental Coal-Titmouse.

111 *Parus ater insularis* (Hellmayr).
Orn. Jahrb., XIII, p. 36, 1902.
Higara. (Japanese Coal-Titmouse.) B.

107 *Parus ater britannicus* Sharpe and Dresser.
Ann. and Mag. Nat. Hist., ser. 4, VIII, p. 437, 1871.
British Coal-Titmouse. B.

108 *Parus ater hibernicus* Ogilvie-Grant.
Bull. B. O. C., XXVII, p. 37, 1910.
Irish Coal-Titmouse. B.

112 *Parus palustris hensoni* Stejneger.
Proc. U.S. Nat. Mus., XV, p. 342, 1892.
Henson-kogara. (Stejneger's Marsh-Titmouse.) B.

109 *Parus palustris dresseri* Stejneger.
Proc. U.S. Nat. Mus., IX, p. 200, 1886.
British Marsh-Titmouse. B.

113 *Parus palustris crassirostris* (Taczanowski).
Bull. Soc. Zool. France, X, p. 470, 1885.
Hashibuto-kogara. (Thick-billed Marsh-Titmouse.) B.

ᵉᵉᵉeeeeeeeeee

114 *Parus atricapillus sachalinensis* Lönnberg.
Journ. Col. Sci. Imp. Univ. Tokyo, XXIII, Art. 14, p. 20, 1908.
Ezo-kogara. (Lönnberg's Willow-Titmouse.) B.

115 *Parus atricapillus restrictus* Hellmayr.
Orn. Jahrb., 1900, p. 215.
Kogara. (Japanese Willow-Titmouse.) B.

110 *Parus atricapillus borealis* Selys-Longchamps.
Bull. Ac. Bruxelles, X, 2, p. 28, 1843.
Northern Willow-Titmouse. (Exc.v.)

111 *Parus atricapillus kleinschmidti* Hellmayr.
Orn. Jahrb., XI, p. 212, 1900.
British Willow-Titmouse. B.

112 *Parus cœruleus cœruleus* L.
Syst. Nat., ed. X, p. 190, 1758.
Continental Blue Titmouse.

113 *Parus cœruleus obscurus* Prazak.
Orn. Jahrb., V, p. 246, 1894.
British Blue Titmouse. B.

114 *Parus cristatus cristatus* L.
Syst. Nat., ed. X, p. 189, 1758.
Northern Crested Titmouse. (R.v.)

115 *Parus cristatus scoticus* (Prazak).
Journ. f. Orn., p. 347, 1897.
Scottish Crested Titmouse. B.

116 *Parus cristatus mitratus* C. L. Brehm.
Handb. Naturg. Vög. Deutschl., p. 467, 1831.
Central European Crested Titmouse. (Exc.v.)

57 Sittiparus Selys-Longchamps (1884)

116 *Sittiparus varius varius* Temminck and Schlegel.
Siebold's Fauna Japon., Aves, p. 71, pl. 35, 1848.
Yamagara. (Varied Titmouse.) B.

117 *Sittiparus varius namiyei* Kuroda.
Dobutsugaku Zasshi (Tokio Zool. Mag.), XXX, p. 322, 1918.
Namiye-yamagara. (Namiye's Varied Titmouse.) B.

118 *Sittiparus varius ijimae* (Kuroda).
Ibis, 1922, p. 98.
Tsushima-yamagara. (Tsushima Varied Titmouse.) B.

119 *Sittiparus varius sunsunpi* Kuroda.
Tokio Zool. Mag., XXXI, p. 231, 1919.
Tane-yamagara. (Tanegashima Varied Titmouse.) B.

120 *Sittiparus varius yakushimensis* Kuroda.
Tokio Zool. Mag., XXXI, p. 232, 1919.
    Yakushima-yamagara. (Yakushima Varied Titmouse.) B.

121 *Sittiparus varius owstoni* Ijima.
Dobutsugaku Zasshi, Yokohama, No. 62, 1893.
    Owston-yamagara. (Ijima's Varied Titmouse.) B.

58 Ægithalos Hermann (1804)

122 *Ægithalos caudatus caudatus* (L.).
Syst. Nat., ed. X, p. 190, 1758.
    Shima-enaga. (Northern Long-tailed Titmouse.) B.

123 *Ægithalos caudatus trivirgatus* (Temminck and Schlegel).
Fauna Japon., Aves, p. 71, pl. 34, 1848.
    Enaga. (Japanese Long-tailed Titmouse.) B.

124 *Ægithalos caudatus kiusiuensis* Kuroda.
Auk, 1923, p. 313.
    Kiusiu-enaga. (Kiusiu Long-tailed Titmouse.) B.

117 *Ægithalos caudatus caudatus* (L.).
Syst. Nat., ed. X, p. 190, 1758.
    Northern Long-tailed Titmouse. (R.v.)

118 *Ægithalos caudatus roseus* (Blyth).
In Gilb. White's Nat. Hist. Selborne, p. 111, 1836.
    British Long-tailed Titmouse. B.

59 Anthoscopus Cabanis (1851)

125 *Anthoscopus pendulinus consobrinus* (Swinhoe).
Proc. Zool. Soc. London, 1870, p. 133.
    Swinhoe-gara. (Chinese Penduline Titmouse.) (Exc.v.)

Family LANIIDÆ

60 Lanius Linnæus (1758)

126 *Lanius sphenocercus sphenocercus* Cabanis.
Journ. f. Orn., 1873, p. 76.
    O-kara-mozu. (Long-tailed Grey Shrike.) (Exc.v.)

127 *Lanius excubitor mollis* Eversmann.
Bull. Soc. Imp. Nat. Moscou, XXVI, p. 498, 1853.
    O-mozu. (Mongolian Grey Shrike.)

128 *Lanius excubitor bianchii* Hartert.
Vogel. Pal. Fauna, I, p. 424, 1907.
    Karafuto-omozu. (Saghalien Island Grey Shrike.) B.

119 *Lanius excubitor excubitor* L.
Syst. Nat., ed. X, p. 94, 1758.
    Great Grey Shrike.

120 *Lanius excubitor meridionalis* Temminck.
Man. d'Orn., ed. II, 1, p. 143, 1820.
    South European Grey Shrike. (Exc.v.)

121 *Lanius minor* Gmelin.
Syst. Nat., I, p. 308, 1788.
Lesser Grey Shrike. (R.v.)

122 *Lanius collurio collurio* L.
Syst. Nat., ed. X, p. 94, 1758.
Red-backed Shrike. B.

129 *Lanius tigrinus* Drapiez.
Dict. Class. Hist. Nat., XIII, p. 523, 1828.
Chigo-mozu. (Thick-billed Shrike.)
B.

130 *Lanius cristatus superciliosus* Latham.
Ind. Orn. Suppl., p. xx, 1801.
Aka-mozu. (Japanese Red-tailed
Shrike.) B.

131 *Lanius bucephalus* Temminck and
Schlegel.
Siebold's Fauna Japonica, Aves, p. 39, pl. 14,
1844.
Mozu. (Bull-headed Shrike.) B.

123 *Lanius senator senator* L.
Syst. Nat., ed. X, p. 94, 1758.
Woodchat Shrike. (R.v.)

124 *Lanius senator badius* Hartlaub.
Journ. f. Orn., 1854, p. 100.
Corsican Woodchat Shrike. (Exc.v.)

125 *Lanius nubicus* Lichtenstein.
Verz. Doubl. Mus. Berlin, p. 47, 1823.
Masked Shrike. (Exc.v.)

Family BOMBYCILLIDÆ

61 Bombycilla Vieillot (1807)

132 *Bombycilla garrulus centralasiae* Pol-
jakow.
Mess. Orn. (Orn. Mitt.), 1915, pp. 137, 138.
Ki-renjaku. (Eastern Waxwing.)

126 *Bombycilla garrulus garrulus* (L.).
Syst. Nat., ed. X, p. 95, 1758.
Waxwing.

133 *Bombycilla japonica* Siebold.
De hist. nat. in Japon. statu etc., p. 13, 1824.
Hi-renjaku. (Japanese Waxwing.)

Family SYLVIIDÆ

62 Phylloscopus Boie (1826)

134 *Phylloscopus tenellipes* Swinhoe.
Ibis, 1860, p. 53.
Ezo-mushikui. (Pale-legged Willow-
Warbler.) B.

135 *Phylloscopus borealis borealis* (Blasius).
Naumannia, p. 313, 1858.
Komushikui. (Eversmann's Warbler.) B.

136 *Phylloscopus borealis xanthodryas* Swinhoe.
Proc. Zool. Soc. London, 1863, p. 296.
Meboso. (Swinhoe's Willow-Warbler.) B.

137 *Phylloscopus occipitalis coronata* (Temminck and Schlegel).
In Siebold's Fauna Japonica, Aves, p. 48, pl. 18, 1847.
Sendai-mushikui. (Temminck's Crowned Willow-Warbler.) B.

138 *Phylloscopus ijimae* (Stejneger).
Proc. U.S. Nat. Mus., XV, p. 372, 1893.
Ijima-meboso. (Seven Island Stejneger's Willow-Warbler.) B.

127 *Phylloscopus trochilus trochilus* (L.).
Syst. Nat., ed. X, p. 188, 1758.
Willow-Warbler. B.

128 *Phylloscopus trochilus eversmanni* (Bonaparte).
Consp. Gen. Av., I, p. 289, 1850.
Northern Willow-Warbler.

129 *Phylloscopus borealis borealis* (Blasius).
Naumannia, p. 313, 1858.
Eversmann's Warbler. (R.v.)

130 *Phylloscopus sibilatrix sibilatrix* (Bechstein).
Naturforscher, XXVII, p. 47, 1793.
Wood-Warbler. B.

131 *Phylloscopus collybita collybita* (Vieillot).
Nouv. Dict. d'Hist. Nat., nouv. éd., XI, p. 235, 1817.
Chiffchaff. B.

132 *Phylloscopus collybita abietinus* (Nilsson).
Kgl. Vet. Akad. Handl., p. 115, 1819.
Scandinavian Chiffchaff. (R.v.)

133 *Phylloscopus collybita tristis* Blyth.
Journ. As. Soc. Bengal, XII, p. 966, 1843.
Siberian Chiffchaff. (R.v.)

134 *Phylloscopus nitidus viridanus* Blyth.
Journ. As. Soc. Bengal, XII, p. 967, 1843.
Greenish Warbler. (Exc.v.)

135 *Phylloscopus humei præmium* Mathews and Iredale.
Austral. Av. Rec., III, p. 44, 1915.
Yellow-browed Warbler.

139 *Phylloscopus proregulus proregulus* (Pallas).
Zoogr. Rosso-Asiat., I, p. 499, 1827.
Karafuto-mushikui. (Pallas's Warbler.) B.

136 *Phylloscopus proregulus proregulus* (Pallas).
Zoogr. Rosso-Asiat., I, p. 499, 1827.
Pallas's Warbler. (Exc.v.)

137 *Phylloscopus fuscatus* (Blyth).
Journ. As. Soc. Bengal, XI, p. 113, 1842.
Dusky Warbler. (Exc.v.)

63 Herbivocula Swinhoe (1871)

140 *Herbivocula schwarzi* (Radde).
Reise Süden v. O. Sibirien, II, p. 260, pl. ix, 1863.
Karafuto-mujisekka. (Radde's Bush-Warbler.) B.

138 *Herbivocula schwarzi* (Radde).
Reise Süden v. O. Sibirien, II, p. 260, pl. ix, 1863.
Radde's Bush-Warbler. (Exc.v.)

64 Horeites Hodgson (1845)

141 *Horeites cantans cantans* (Temminck and Schlegel).
Siebold's Fauna Japonica, Aves, p. 51, pl. 19, 1847.
Uguisu. (Large Japanese Bush-Warbler.) B.

142 *Horeites cantans ijimae* (Kuroda).
Annot. Zool. Japon., X, p. 117, 1922.
Ijima-uguisu. (Kuroda's Bush-Warbler.) B.

143 *Horeites cantans medius* (Momiyama).
Tokyo Zool. Mag., Feb. 1924.
Hachijo-uguisu. (Hachijo Island Bush-Warbler.) B.

65 Cettia Bonaparte (1838)

139 *Cettia cetti cetti* (Temminck).
Man. d'Orn., 2nd ed., I, p. 194, 1820.
Cetti's Warbler. (Exc.v.)

66 Lusciniola Gray (1841)

140 *Lusciniola melanopogon melanopogon* (Temminck).
Pl. Color., 245, fig. 2, 1823.
Moustached Warbler. (Exc.v.)

67 Locustella Kaup (1829)

141 *Locustella nævia nævia* (Boddaert).
Tabl. Pl. Enl., p. 35, 1783.
Grasshopper-Warbler. B.

144 *Locustella fasciolatus* (Gray).
Proc. Zool. Soc. London, 1860, p. 349.
Yezo-senniu. (Gray's Grasshopper-Warbler.) B.

145 *Locustella ochotensis ochotensis* (Middendorff).
Sibir. Reise, II, 2, p. 185, pl. xvi, fig. 7, 1853.
Shima-senniu. (Middendorff's Grasshopper-Warbler.) B.

146 *Locustella ochotensis pleskei* (Taczanowski).
Proc. Zool. Soc. London, 1889, p. 620.
Uchiyama-senniu. (Hondo Grasshopper-Warbler.) B.

147 *Locustella lanceolata* (Temminck).
Man. d'Orn., 2nd ed., IV, p. 614, 1840.
Makino-senniu. (Lanceolated War-
bler.) B.

142 *Locustella lanceolata* (Temminck).
Man. d'Orn., 2nd ed., IV, p. 614, 1840.
Lanceolated Warbler. (R.v.)

143 *Locustella certhiola* (Pallas).
Zoogr. Rosso-Asiat., I, p. 509, 1827.
Pallas's Grasshopper-Warbler. (Exc.v.)

144 *Locustella luscinioides luscinioides*
(Savi).
Nuovo Giorn. Letter., VII, p. 341, 1824.
Savi's Warbler.

68 Acrocephalus Naumann (1811)

145 *Acrocephalus arundinaceus arundina-
ceus* (L.).
Syst. Nat., ed. X, p. 170, 1758.
Great Reed-Warbler.

148 *Acrocephalus arundinaceus orientalis*
(Temminck and Schlegel).
Siebold's Fauna Japon., Aves, p. 50, pl. XX *b*,
1847.
O-yoshikiri. (Eastern Great Reed-
Warbler.) B.

146 *Acrocephalus arundinaceus orientalis*
(Temminck and Schlegel).
Siebold's Fauna Japon., Aves, p. 50, pl. XX *b*,
1847.
Eastern Great Reed-Warbler.
(Exc.v.)

149 *Acrocephalus bistrigiceps* Swinhoe.
Ibis, 1860, p. 51.
Koyoshikiri. (Schrenck's Reed-
Warbler.) B.

147 *Acrocephalus scirpaceus scirpaceus* (Her-
mann).
Observ. Zool., p. 202, 1804.
Reed-Warbler. B.

148 *Acrocephalus palustris* (Bechstein).
Orn. Taschenb., p. 186, 1803.
Marsh-Warbler. B.

149 *Acrocephalus dumetorum* Blyth.
Journ. As. Soc. Bengal, XVIII, p. 815, 1849.
Blyth's Reed-Warbler. (Exc.v.)

150 *Acrocephalus schœnobænus* (L.).
Syst. Nat., ed. X, p. 184, 1758.
Sedge-Warbler. B.

151 *Acrocephalus paludicola* (Vieillot).
Nouv. Dict. d'Hist. Nat., nouv. éd., XI, p. 202, 1817.
Aquatic Warbler.

69 Hippolais Brehm (1828)

152 *Hippolais icterina* (Vieillot).
Nouv. Dict. d'Hist. Nat., nouv. éd., XI, p. 194, 1817.
Icterine Warbler.

153 *Hippolais polyglotta* (Vieillot).
Nouv. Dict. d'Hist. Nat., nouv. éd., XI, p. 200, 1817.
Melodious Warbler. (Exc.v.)

154 *Hippolais pallida elæica* (Lindermayer).
Isis, 1843, pp. 342, 343.
Olivaceous Warbler. (Exc.v.)

70 Sylvia Scopoli (1769)

155 *Sylvia communis communis* Latham.
Gen. Syn. Suppl., I, p. 287, 1787.
White-throat. B.

156 *Sylvia curruca curruca* (L.).
Syst. Nat., ed. X, p. 184, 1758.
Lesser Whitethroat. B.

157 *Sylvia curruca affinis* Blyth.
Journ. As. Soc. Bengal, XIV, p. 564, 1845.
Siberian Lesser Whitethroat. (Exc.v.)

158 *Sylvia rüppelli* Temminck.
Pl. Color., 245, fig. 1, 1823.
Rüppell's Warbler. (Exc.v.)

159 *Sylvia borin* (Boddaert).
Tabl. Pl. Enl., p. 35, 1783.
Garden-Warbler. B.

160 *Sylvia atricapilla atricapilla* (L.).
Syst. Nat., ed. X, p. 187, 1758.
Blackcap. B.

161 *Sylvia melanocephala melanocephala* (Gmelin).
Syst. Nat., I, II, p. 970, 1789.
Sardinian Warbler. (Exc.v.)

162 *Sylvia hortensis hortensis* (Gmelin).
Syst. Nat., I, II, p. 955, 1789.
Orphean Warbler. (Exc.v.)

163 *Sylvia nisoria nisoria* (Bechstein).
Gem. Naturg. Deutschl., IV, p. 580, 1795.
Barred Warbler.

164 *Sylvia cantillans cantillans* (Pallas).
In Vroeg's Cat. Verzam. Vogelen, etc., Adumbratiuncula, p. 4, 1764.
Subalpine Warbler. (Exc.v.)

71 **Melizophilus** Forster (1817)

165 *Melizophilus undatus dartfordiensis* (Latham).
Gen. Syn. Suppl., I, p. 287, 1787.
Dartford Warbler. B.

72 **Agrobates** Swainson (1837)

166 *Agrobates galactotes galactotes* (Temminck).
Man. d'Orn., ed. II, I, p. 182, 1820.
Rufous Warbler. (Exc.v.)

167 *Agrobates galactotes syriacus* (Hemprich and Ehrenberg).
Symb. Phys., fol. bb, 1833.
Brown-backed Warbler. (Exc.v.)

73 **Bradypterus** Swainson (1837)

150 *Bradypterus pryeri pryeri* (Seebohm).
Ibis, 1884, p. 40.
O-sekka. (Pryer's Grass Warbler.)

74 **Cisticola** Kaup (1829)

151 *Cisticola cisticola brunniceps* (Temminck and Schlegel).
In Siebold's Fauna Japon., Aves, p. 134, pl. 20, 1850.
Sekka. (Japanese Fantail Warbler.)
B.

152 *Cisticola cisticola djadja* Momiyama.
Dobutsu. Zasshi (Tokyo Zool. Mag.), p. 408, 1923.
Shima-sekka. (Seven Island Fantail Warbler.) B.

Family TURDIDÆ

75 **Turdus** Linnæus (1758)

168 *Turdus viscivorus viscivorus* L.
Syst. Nat., ed. x, p. 168, 1758.
Mistle-Thrush. B.

169 *Turdus philomelos philomelos* Brehm.
Handb. Naturg. Vög. Deutschl., p. 382, 1831.
Continental Song-Thrush.

170 *Turdus philomelos clarkei* Hartert.
Bull. B. O. C., XXIII, p. 54, 1909
British Song-Thrush. B.

171 *Turdus philomelos hebridensis* Clarke.
Scott. Nat., p. 53, 1913.
Hebridean Song-Thrush. B.

172 *Turdus musicus* L.
Syst. Nat., ed. x, p. 169, 1758.
Redwing.

173 *Turdus pilaris* L.
Syst. Nat., ed. x, p. 168, 1758.
Fieldfare.

153 *Turdus cardis* Temminck.
Pl. Color. 518, 1830.
Kuro-tsugumi. (Grey Japanese Ouzel.) B.

154 *Turdus eunomus* Temminck.
Pl. Color. 514, 1830.
Tsugumi. (Dusky Thrush.)

174 *Turdus eunomus* Temminck.
Pl. Color. 514, 1830.
Dusky Thrush. (Exc.v.)

155 *Turdus naumanni* Temminck.
Man. d'Orn., ed. II, I, p. 170, 1820.
Hachijo-tsugumi. (Red-tailed Ouzel.) (R.v.)

175 *Turdus ruficollis atrogularis* Temminck.
Man. d'Orn., ed. II, I, p. 169, 1820.
Black-throated Thrush. (Exc.v.)

156 *Turdus hortulorum* Sclater.
Ibis, 1863, p. 196.
Kara-akahara. (Swinhoe's Thrush.) (R.v.)

157 *Turdus obscurus* Gmelin.
Syst. Nat., I, II, p. 816, 1789.
Mamichajinai. (Eyebrowed Ouzel.)

158 *Turdus pallidus* Gmelin.
Syst. Nat., I, II, p. 815, 1789.
Shirohara. (Pale Thrush.) B.

159 *Turdus chrysolaus* Temminck.
Pl. Color. 537, 1831.
Akahara. (Japanese Brown Thrush.)
B.

160 *Turdus celaenops celaenops* Stejneger.
Science, X, p. 108, 1887 and Proc. U.S. Nat.
Mus., 1887, p. 484.
Akakokko. (Seven Island Ouzel.) B.

161 *Turdus celaenops kurodai* (Momiyama).
Dobutsu. Zasshi (Tokyo Zool. Mag.), p. 404,
1923.
Hachijo-kokko. (Seven Island
Thrush.) B.

162 *Turdus celaenops yakushimensis* (Oga-
wa).
Annot. Zool. Japon., v, p. 180, 1905.
Yakushima-akakokko. (Yakushima
Thrush.) B.

176 *Turdus merula merula* L.
Syst. Nat., ed. X, p. 170, 1758.
Blackbird. B.

177 *Turdus torquatus torquatus* L.
Syst. Nat., ed. X, p. 170, 1758.
Ring-Ouzel. B.

178 *Turdus torquatus alpestris* (C. L.
Brehm).
Isis, 1828, p. 1281.
Alpine Ring-Ouzel. (Exc.v.)

163 *Turdus aureus* Holandre.
Fauna dép. Moselle, in Ann. Moselle, p. 60,
1825.
Tora-tsugumi. (White's Thrush.) B.

179 *Turdus aureus* Holandre.
Fauna dép. Moselle, in Ann. Moselle, p. 60,
1825.
White's Thrush. (R.v.)

76 Geokichla S. Müller 1835 (ex Boie)

164 *Geocichla sibiricus davisoni* (Hume).
Stray Feathers, v, p. 63, 1877.
Mamijiro. (Davison's Ground
Thrush.) B.

77 Monticola Boie (1822)

180 *Monticola saxatilis* (L.).
Syst. Nat., ed. XII, p. 294, 1766.
Rock-Thrush. (R.v.)

165 *Monticola solitarius magnus* (La
Touche).
Bull. B. O. C., XL, p. 97, 1919.
Iso-hiyodori. (Japanese Blue Rock-
Thrush.) B.

78 Phœnicurus Forster (1817)

181 *Phœnicurus phœnicurus phœnicurus* (L.).
Syst. Nat., ed. x, p. 187, 1758.
Redstart. B.

182 *Phœnicurus ochrurus gibraltariensis* (Gmelin).
Syst. Nat., I, II, p. 987, 1789.
Black Redstart.

166 *Phœnicurus aurorea aurorea* (Pallas).
Reise d. versch. Prov. d. Russ. Reichs, III, p. 695, 1776.
Jobitaki. (Daurian Redstart.) B.

79 Erithacus Cuvier (1800)

183 *Erithacus rubecula rubecula* (L.).
Syst. Nat., ed. x, p. 188, 1758.
Continental Robin.

184 *Erithacus rubecula melophilus* (Hartert).
Nov. Zool., VIII, p. 317, 1901.
British Robin. B.

167 *Erithacus akahige akahige* (Temminck).
Pl. Color. 571, 1824.
Komadori. (Japanese Robin.) B.

168 *Erithacus akahige spectatoris* Momi-yama.
Dobutsu. Zasshi (Tokyo Zool. Mag.), p. 403, 1923.
Shichito-komadori. (Seven Island Robin.) B.

169 *Erithacus akahige tanensis* Kuroda.
Bull. B. O. C., XLIII, p. 106, Jan. 1923.
Tane-komadori. (Kuroda's Robin.) B.

170 *Erithacus sibilans* (Swinhoe).
Proc. Zool. Soc. Lond., 1863, p. 292.
Shimagoma. (Swinhoe's Robin.) B.

80 Luscinia Forster (1817)

185 *Luscinia megarhyncha megarhyncha* C. L. Brehm.
Handb. Naturg. Vög. Deutschl., p. 356, 1831.
Nightingale. B.

186 *Luscinia luscinia* (L.).
Syst. Nat., ed. x, p. 184, 1758.
Thrush-Nightingale. (Exc.v.)

81 **Cyanosylvia** Brehm (1828)

171 *Cyanosylvia svecica robusta* (Buturlin).
Psoveia i Rusheinaia Okhota (Jagen and Schieben), March 1907, No. 6, and Orn. Monatsber., May 1907, p. 79.
Ogawa-komadori. (East Siberian Bluethroat.) (Exc.v.)

187 *Cyanosylvia svecica svecica* (L.).
Syst. Nat., ed. x, p. 187, 1758.
Lapland Bluethroat. (Exc.v.)

188 *Cyanosylvia svecica gaetkei* (Kleinschmidt).
Journ. f. Orn., 1904, p. 302.
Norwegian Bluethroat.

189 *Cyanosylvia svecica cyanecula* (Wolf).
In Meyer and Wolf's Taschenb. d. deutsch. Vögelk., I, p. 240, 1810.
White-spotted Bluethroat. (R.v.)

82 **Ianthia** Blyth (1847)

172 *Ianthia cyanura* (Pallas).
Reise d. versch. Prov. Russ. Reichs, II, p. 709, 1773.
Ruribitaki. (Siberian Blue-tail.) B.

83 **Saxicola** Bechstein (1802)

173 *Saxicola torquata stejnegeri* (Parrot).
Verh. orn. Ges. Bayern, VIII, p. 124, 1908.
Nobitaki. (Eastern Stonechat.) B.

190 *Saxicola torquata hibernans* (Hartert).
Journ. f. Orn., 1910, p. 173.
British Stonechat. B.

191 *Saxicola torquata indica* (Blyth).
Journ. As. Soc. Bengal, XVI, p. 129, 1847.
Indian Stonechat. (Exc.v.)

192 *Saxicola rubetra rubetra* (L.).
Syst. Nat., ed. x, p. 186, 1758.
Whinchat. B.

84 **Œnanthe** Vieillot (1816)

193 *Œnanthe œnanthe œnanthe* (L.).
Syst. Nat., ed. x, p. 186, 1758.
Wheatear. B.

194 *Œnanthe œnanthe leucorrhoa* (Gmelin).
Syst. Nat., I, II, p. 966, 1789.
Greenland Wheatear.

195 *Œnanthe isabellina* (Cretzschmar).
Atlas zu Rüppells Reise, Vögel, p. 52, pl. 34, b, 1826.
Isabelline Wheatear. (Exc.v.)

196 *Œnanthe hispanica hispanica* (L.).
Syst. Nat., ed. x, p. 186, 1758.
Western Black-eared Wheatear. (R.v.)

197 *Œnanthe hispanica melanoleuca* (Gül-
denstädt).
Nov. Comm. Petrop., XIX, p. 468, pl. 15, 1775.
Eastern Black-eared Wheatear.
(Exc.v.)

198 *Œnanthe deserti homochroa* (Tristram).
Ibis, 1859, p. 59.
Western Desert-Wheatear. (Exc.v.)

199 *Œnanthe deserti atrogularis* (Blyth).
Journ. Asiatic Soc. Bengal, XVI, p. 131, 1847.
Eastern Desert-Wheatear. (Exc.v.)

200 *Œnanthe leucomela leucomela* (Pallas).
Nov. Comm. Petr., XIV, p. 584, pl. 22, fig. 3,
1770.
Pied Wheatear. (Exc.v.)

201 *Œnanthe leucura leucura* (Gmelin).
Syst. Nat., I, II, p. 820, 1789.
Black Wheatear. (Exc.v.)

202 *Œnanthe leucura syenitica* (Heuglin).
Journ. f. Orn., 1869, p. 155.
North African Black Wheatear.
(Exc.v.)

Family PRUNELLIDÆ

85 Prunella Vieillot (1816)

203 *Prunella modularis modularis* (L.).
Syst. Nat., ed. X, p. 184, 1758.
Continental Hedge-Sparrow.

204 *Prunella modularis occidentalis* (Har-
tert).
Brit. Birds, III, p. 313, March 1910.
British Hedge-Sparrow. B.

174 *Prunella collaris erythropygius* (Swinhoe).
Proc. Zool. Soc. London, 1870, p. 124, pl. 9.
Iwa-hibari. (Oriental Alpine Accentor.) B.

205 *Prunella collaris collaris* (Scopoli).
Annus I Historico-Natur., p. 131, 1769.
Alpine Accentor. (R.v.)

175 *Prunella rubidus* (Temminck and
Schlegel).
Siebold's Fauna Japon., Aves, p. 69, pl. 32, 1850.
Kayakuguri. (Japanese Hedge-
Sparrow.) B.

176 *Prunella montanella* (Pallas).
Reise versch. Prov. Russ. Reichs, III, p. 695,
1776.
Yamahibari. (Siberian Hedge-
Sparrow.) (Exc.v.)

Family TROGLODYTIDÆ

86 **Troglodytes** Vieillot (1807)

177 *Troglodytes troglodytes dauricus* Dybowski and Taczanowski.
Bull. Soc. Zool. France, 1884, p. 155.
Karafuto-misosazai. (Daurian Wren.)
B.

178 *Troglodytes troglodytes kurilensis* Stejneger.
Proc. U.S. Nat. Mus., 1888, p. 548.
Chishima-misosazai. (Kurile Island Wren.) B.

179 *Troglodytes troglodytes fumigatus* Temminck.
Man. d'Orn., ed. II, III, p. 161, 1835.
Misosazai. (Japanese Wren.) B.

180 *Troglodytes troglodytes utanoi* Kuroda.
Ibis, 1922, p. 96.
Tsushima-misosazai. (Kuroda's Wren.) B.

181 *Troglodytes troglodytes mosukei* Momiyama.
Dobutsu. Zasshi (Tokyo Zool. Mag.), p. 402, 1923.
Mosuke-misosazai. (Momiyama's Wren.) B.

182 *Troglodytes troglodytes ogawae* Hartert.
Vog. Pal. Fauna, I, p. 784, 1910.
Ogawa-misosazai. (Ogawa's Wren.) B.

206 *Troglodytes troglodytes troglodytes* (L.).
Syst. Nat., ed. x, p. 188, 1758.
Wren. B.

207 *Troglodytes troglodytes hirtensis* Seebohm.
Zoologist, 1884, p. 333.
St. Kilda Wren. B.

208 *Troglodytes troglodytes zetlandicus* Hartert.
Vög. pal. Fauna, I, p. 777, 1910.
Shetland Wren. B.

87 **Cinclus** Borkhausen (1797)

209 *Cinclus cinclus cinclus* (L.).
Syst. Nat., ed. x, p. 168, 1758.
Black-bellied Dipper.

210 *Cinclus cinclus gularis* (Latham).
2nd Suppl. Gen. Synops., p. XL, 1801.
British Dipper. B.

211 *Cinclus cinclus hibernicus* Hartert.
Vög. pal. Fauna, I, p. 790, 1910.
Irish Dipper. B.

183 *Cinclus pallasii pallasii* Temminck.
Man. d'Orn., ed. II, I, p. 177, 1820.
Kawagarasu. (Siberian Black-bellied Dipper.) B.

Family HIRUNDINIDÆ

88 Hirundo Linnæus (1758)

184 *Hirundo rustica gutturalis* Scopoli.
Del. Flor. and Faun. Insubr., II, p. 96, 1786.
Tsubame. (Eastern Chimney-Swallow.) B.

212 *Hirundo rustica rustica* L.
Syst. Nat., ed. x, p. 191, 1758.
Swallow. B.

185 *Hirundo rustica tytleri* Jerdon.
B. India, III, p. 870, 1864.
Akahara-tsubame. (Tytler's Chimney-Swallow.) (R.v.)

186 *Hirundo daurica nipalensis* Hodgson.
Journ. As. Soc. Bengal, v, 1836, p. 780.
Koshiaka-tsubame. (Hodgson's Mosque Swallow.) B.

213 *Hirundo daurica rufula* Temminck.
Man. d'Orn., ed. II, III, p. 298, 1835.
Red-rumped Swallow. (Exc.v.)

89 Delichon Moore (1854)

187 *Delichon urbica whiteleyi* (Swinhoe).
Proc. Zool. Soc. London, 1862, p. 320.
Siberia-iwatsubame. (Pallas's Martin.)

214 *Delichon urbica urbica* (L.).
Syst. Nat., ed. x, p. 192, 1758.
Martin. B.

188 *Delichon urbica dasypus* (Bonaparte).
Consp. Av., I, p. 343, 1850.
Iwa-tsubame. (Blakiston's Martin.) B.

90 Riparia Forster (1817)

189 *Riparia riparia ijimae* (Lönnberg).
Journ. College Science Tokyo, XXIII, Art. 14, p. 38, 1908.
Shodo-tsubame. (Japanese Sand-Martin.) B.

215 *Riparia riparia riparia* (L.).
Syst. Nat., ed. x, p. 192, 1758.
Sand-Martin. B.

Family CAMPEPHAGIDÆ

91 Pericrocotus Boie (1826)

190 *Pericrocotus cinereus* Lafresnaye.
Rev. Zool., 1845, p. 94.
Sanshokui. (Ashy Minivet.) B.

191 *Pericrocotus tegimae* Stejneger.
Proc. U.S. Nat. Mus., IX, 1886, p. 648.
Riukiu-sanshokui. (Loo-choo Island Minivet.) B.

Family PITTIDÆ

92 Pitta Vieillot (1816)

192 *Pitta nympha* Temminck and Schlegel.
Siebold's Fauna Japon., Aves, p. 135, Suppl. pl. A, 1850.
Yairocho. (Fairy Pitta.) (R.v.)

Order PICI

Family PICIDÆ

93 Picus Linnæus (1758)

216 *Picus viridis virescens* (C. L. Brehm).
Handb. Naturg. Vog. Deutschl., p. 199, 1831.
Green Woodpecker. B.

193 *Picus canus jessoensis* Stejneger.
Proc. U.S. Nat. Mus., IX, p. 106, 1886.
Yamagera. (Japanese Grey-headed
Green Woodpecker.) B.

194 *Picus awokera awokera* Temminck.
Pl. Color., 585, 1826.
Awogera. (Japanese Green Wood-
pecker.) B.

195 *Picus awokera horii* Takatsukasa.
Dobutsu. Zasshi (Tokyo Zool. Mag.), XXX,
p. 422, 1918.
Kagoshima-awogera. (Kiusiu Green
Woodpecker.) B.

196 *Picus awokera takatsukasae* Kuroda.
Auk, 1921, p. 576.
Tane-awogera. (Kuroda's Green
Woodpecker.) B.

94 Dryobates Boie (1826)

197 *Dryobates major tscherskii* (Buturlin).
Nascha Okhota (Unser Sport), St Petersburg,
Juli 1910, p. 53.
Kita-akagera. (Buturlin's Great
Spotted Woodpecker.) B.

217 *Dryobates major major* (L.).
Syst. Nat., ed. X, p. 114, 1758.
Northern Great Spotted Wood-
pecker.

198 *Dryobates major japonicus* (Seebohm).
Ibis, 1883, p. 24.
Yezo-akagera. (Japanese Great
Spotted Woodpecker.) B.

218 *Dryobates major anglicus* (Hartert).
Nov. Zool., 1900, p. 528.
British Great Spotted Woodpecker.
B.

199 *Dryobates major hondoensis* Kuroda.
Auk, 1921, p. 577.
Akagera. (Kuroda's Great Spotted
Woodpecker.) B.

200 *Dryobates leucotos uralensis* (Malherbe).
Monogr. Picidés, I, p. 92, pl. 23, figs. 4, 5, 1861.
Chosen-o-akagera. (Ural White-
backed Woodpecker.) B.

201 *Dryobates leucotos subcirris* Stejneger.
Proc. U.S. Nat. Mus., IX, p. 113, 1886.
Yezo-o-akagera. (Yezo White-
backed Woodpecker.) B.

202 *Dryobates leucotos stejnegeri* Kuroda.
Auk, 1921, p. 579.
O-akagera. (Japanese White-backed Woodpecker.) B.

203 *Dryobates leucotos intermedius* Kuroda.
Auk, 1921, p. 580.
Kansai-o-akagera. (Kuroda's White-backed Woodpecker.) B.

204 *Dryobates leucotos namiyei* Stejneger.
Proc. U.S. Nat. Mus., IX, p. 116, 1886.
Namiye-gera. (Stejneger's Woodpecker.) B.

205 *Dryobates minor minutillus* (Buturlin).
Annuaire Mus. Zool. Petersbourg, XIII, p. 246, 1908.
Ko-akagera. (Japanese Lesser Spotted Woodpecker.) B.

206 *Dryobates kizuki ijimae* Takatsukasa.
Dobutsu. Zasshi (Tokyo Zool. Mag.), 1921, Dr Ijima's Memorial Ed.
Ijima-kogera. (Saghalien Island Pigmy Woodpecker.) B.

207 *Dryobates kizuki seebohmi* Hargitt.
Ibis, 1884, p. 100.
Yezo-kogera. (Hargitt's Pigmy Woodpecker.) B.

208 *Dryobates kizuki nippon* Kuroda.
Ibis, 1922, p. 88.
Kogera. (Hondo Pigmy Woodpecker.) B.

209 *Dryobates kizuki shikokuensis* Kuroda.
Annot. Zool. Japon., X, p. 115, 1922.
Shikoku-kogera. (Kuroda's Pigmy Woodpecker.) B.

210 *Dryobates kizuki kizuki* (Temminck).
Pl. Color., IV, livr. 99, nec. pl. 585, 1836.
Kiushiu-kogera. (Temminck's Pigmy Woodpecker.) B.

211 *Dryobates kizuki harterti* Kuroda.
Bull. B.O.C., XLIII, p. 108, 1923.
Yakushima-kogera. (Yakushima Pigmy Woodpecker.) B.

212 *Dryobates kizuki matsudairai* Kuroda.
Auk, 1921, p. 576.
Miyake-kogera. (Japanese Pigmy Woodpecker.) B.

219 *Dryobates minor comminutus* (Hartert).
Brit. B., I, p. 221, 1907.
British Lesser Spotted Woodpecker. B.

95 Picoides Lacépède (1799)

213 *Picoides tridactylus sakhalinensis* Buturlin.
Orn. Monatsber., 1907, pp. 10, 11.
  Miyubigera. (Japanese Three-toed Woodpecker.) B.

96 Dryocopus Boie (1826)

214 *Dryocopus martius silvifragus* Riley.
Proc. Biol. Soc. Washington, 28, p. 162, 1915.
  Kumagera. (Japanese Great Black Woodpecker.) B.

97 Jynx Linnæus (1758)

215 *Jynx torquilla japonica* Bonaparte.
Consp. Avium, I, p. 112, 1850.
  Arisui. (Japanese Wryneck.) B.

220 *Jynx torquilla torquilla* L.
Syst. Nat., ed. x, p. 112, 1758.
  Wryneck. B.

216 *Jynx torquilla hokkaidi* (Kuroda).
Auk, 1921, p. 582.
  Yezo-arisui. (Kuroda's Wryneck.)

Order CUCULI

Family CUCULIDÆ

98 Cuculus Linnæus (1758)

217 *Cuculus canorus telephonus* Heine.
Journ. f. Orn., 1863, p. 352.
  Kakko. (Asiatic Cuckoo.) B.

221 *Cuculus canorus canorus* L.
Syst. Nat., ed. x, p. 110, 1758.
  Cuckoo. B.

218 *Cuculus optatus optatus* Gould.
Proc. Zool. Soc. London, part XIII, 1845, p. 18.
  Tsutsudori. (Himalayan Cuckoo.) B.

219 *Cuculus poliocephalus poliocephalus* Latham.
Index Orn. I, p. 214, 1790.
  Hototogisu. (Little Cuckoo.) B.

220 *Cuculus micropterus micropterus* Gould.
Proc. Zool. Soc. London, 1837, p. 137.
  Seguro-kakko. (Indian Cuckoo.) (Exc.v.)

99 Hierococcyx S. Müller (1839–44)

221 *Hierococcyx fugax nisicolor* (Blyth).
Journ. As. Soc. Bengal, XII, p. 943, 1843.
  Jiuichi. (Amoor Cuckoo.) B.

222 *Hierococcyx sparverioides* (Vigors).
Proc. Committee Zool. Soc. London, part I, p. 173, 1832.
  O-jiuichi. (Hawk Cuckoo.) (Exc.v.)

100 **Clamator** Kaup (1829)

    222 *Clamator glandarius* (L.).
      Syst. Nat., ed. x, p. 111, 1758.
        Great Spotted Cuckoo. (Exc.v.)

101 **Coccyzus** Vieillot (1816)

    223 *Coccyzus americanus americanus* (L.).
      Syst. Nat., ed. x, p. 111, 1758.
        American Yellow-billed Cuckoo.
        (R.v.)

    224 *Coccyzus erythropthalmus* (Wilson).
      Amer. Orn., iv, p. 16, pl. 28, fig. 2, 1811.
        American Black-billed Cuckoo.
        (Exc.v.)

Order CYPSELI

Family CYPSELIDÆ

102 **Apus** Scopoli (1777)

    225 *Apus apus apus* (L.).
      Syst. Nat., ed. x, p. 192, 1758.
        Swift. B.

    226 *Apus melba melba* (L.).
      Syst. Nat., ed. x, p. 192, 1758.
        Alpine Swift. (R.v.)

223 *Apus pacificus pacificus* (Latham).
  Index Orn., Suppl., p. lviii, 1801.
    Amatsubame. (White-rumped Swift.)
    B.

103 **Hirundapus** Hodgson (1836)

224 *Hirundapus caudacuta caudacuta*
(Latham).
  Index Orn., Suppl., p. lvii, 1801.
    Hariwo-amatsubame. (Needle-tailed
  Swift.) B.

    227 *Hirundapus caudacuta caudacuta*
    (Latham).
      Index Orn., Suppl., p. lvii, 1801.
        Needle-tailed Swift. (Exc.v.)

Order CAPRIMULGI

Family CAPRIMULGIDÆ

104 **Caprimulgus** Linnæus (1758)

    228 *Caprimulgus europæus europæus* (L.).
      Syst. Nat., ed. x, p. 193, 1758.
        Nightjar. B.

225 *Caprimulgus indicus jotaka* Temminck
and Schlegel.
  Siebold's Fauna Japonica, Aves, p. 37, pl. 12,
  1847.
    Yotaka. (Japanese Nightjar.) B.

229 *Caprimulgus ruficollis desertorum* Erlanger.
Journ. f. Orn., 1899, p. 521, pl. xi.
     Algerian Red-necked Nightjar. (Exc.v.)

230 *Caprimulgus ægyptius ægyptius* Lichtenstein.
Verz. Doubl., p. 59, 1823.
     Egyptian Nightjar. (Exc.v.)

Order MEROPES

Family MEROPIDÆ

105 Merops Linnæus (1758)

231 *Merops apiaster* L.
Syst. Nat., ed. x, p. 117, 1758.
     Bee-eater. (R.v.)

Order UPUPÆ

Family UPUPIDÆ

106 Upupa Linnæus (1758)

226 *Upupa epops saturata* Lönnberg.
Arkiv för Zoologi, v, no. 9, p. 29, 1909.
     Yatsugashira. (Oriental Hoopoe.)

232 *Upupa epops epops* L.
Syst. Nat., ed. x, p. 117, 1758.
     Hoopoe.

Order HALCYONES

Family ALCEDINIDÆ

107 Alcedo Linnæus (1758)

227 *Alcedo atthis japonica* Bonaparte.
Consp. Vol. Anis., p. 10, 1854.
     Kawasemi. (Japanese Kingfisher.) B.

233 *Alcedo atthis ispida* L.
Syst. Nat., ed. x, p. 115, 1758.
     Kingfisher. B.

108 Ceryle Boie (1828)

228 *Ceryle lugubris lugubris* (Temminck).
Pl. Color., 548, 1834.
     Yamasemi. (Oriental Spotted Kingfisher.) B.

109 Halcyon Swainson (1820)

229 *Halcyon pileata* (Boddaert).
Tabl. Pl. Enl., p. 41, 1783.
     Yama-shobin. (Black-capped Kingfisher.)

230 *Halcyon coromanda major* Temminck and Schlegel.
Siebold's Fauna Japon., Aves, pp. 75, 76, pl. 39, 1842.
     Aka-shobin. (Japanese Ruddy Kingfisher.) B.

Order CORACIÆ

Family CORACIIDÆ

110 Coracias Linnæus (1758)

234 *Coracias garrulus garrulus* L.
Syst. Nat., ed. x, p. 107, 1758.
Roller.

111 Eurystomus Vieillot (1816)

231 *Eurystomus orientalis calonyx* Sharpe.
Proc. Zool. Soc. London, 1890, p. 551.
Bupposo. (Sharpe's Broad-billed
Roller.) B.

Order STRIGES

Family STRIGIDÆ

112 Tyto Billberg (1828)

235 *Tyto alba alba* (Scopoli).
Annus I, Historico-Natur., p. 21, 1769.
White-breasted Barn-Owl. B.

236 *Tyto alba guttata* (C. L. Brehm).
Handb. Naturg. Vög. Deutschl., p. 106, 1831.
Dark-breasted Barn-Owl.

113 Asio Brisson (1760)

232 *Asio otus otus* (L.).
Syst. Nat., ed. x, p. 92, 1758.
Torafu-zuku. (Long-eared Owl.) B.

233 *Asio flammeus flammeus*(Pontoppidan).
Danske Atlas, I, p. 617, pl. xxv, 1763.
Komimizuku. (Short-eared Owl.) B.

234 *Asio flammeus leucopsis* (Brehm).
Vogelfang, p. 413, 1855.
Sameiro-komimizuku. (Japanese
Short-eared Owl.) (Exc.v.)

237 *Asio otus otus* (L.).
Syst. Nat., ed. x, p. 92, 1758.
Long-eared Owl. B.

238 *Asio flammeus flammeus*(Pontoppidan).
Danske Atlas, I, p. 617, pl. xxv, 1763.
Short-eared Owl. B.

114 Strix Linnæus (1758)

239 *Strix aluco sylvatica* Shaw.
Gen. Zool., VII, p. 253, 1809.
British Tawny Owl. B.

235 *Strix nebulosa sakhalinensis* (Buturlin).
Journ. f. Orn., Juli 1907, pp. 332, 334.
Karafuto-fukuro. (Saghalien Island
Great Gray Owl.) B.

236 *Strix uralensis nikolskii* (Buturlin).
Journ. f. Orn., 1907, pp. 333, 334.
    Shiberia-fukuro. (Buturlin's Ural Owl.) B.

237 *Strix uralensis japonica* (Clark).
Proc. U.S. Nat. Mus., XXXII, p. 471, 1907.
    Yezo-fukuro. (Clark's Ural Owl.) B.

238 *Strix uralensis hondoensis* (Clark).
Proc. U.S. Nat. Mus., XXXII, p. 472, 1907.
    Fukuro. (Hondo Ural Owl.) B.

239 *Strix uralensis fuscescens* Temminck and Schlegel.
Fauna Japon., Aves, pl. 10, 1845.
    Kiushiu-fukuro. (Kiusiu Ural Owl.) B.

115 Nyctea Stephens (1826)

240 *Nyctea nyctea* (L.).
Syst. Nat., ed. x, p. 93, 1758.
    Shiro-fukuro. (Snowy Owl.)

240 *Nyctea nyctea* (L.).
Syst. Nat., ed. x, p. 93, 1758.
    Snowy Owl.

116 Surnia Duméril (1806)

241 *Surnia ulula pallasi* Buturlin.
Orn. Monatsber., 1907, p. 100.
    Onaga-fukuro. (Siberian Hawk-Owl.) (R.v.)

241 *Surnia ulula ulula* (L.).
Syst. Nat., ed. x, p. 93, 1758.
    European Hawk-Owl. (Exc.v.)

242 *Surnia ulula caparoch* (P. L. S. Muller).
Natursystem, Suppl., p. 69, 1776.
    American Hawk-Owl. (Exc.v.)

117 Otus Pennant (1769)

243 *Otus scops scops* (L.).
Syst. Nat., ed. x, p. 92, 1758.
    Scops-Owl. (R.v.)

242 *Otus bakkamoena semitorques* Temminck and Schlegel.
In Siebold's Fauna Japon., Aves, p. 95, pl. 8, 1850.
    O-konohazuku. (Feather-toed Scops-Owl.) B.

243 *Otus bakkamoena hatchizionis* Momiyama.
Dobutsu. Zasshi (Tokyo Zool. Mag.), p. 400, 1923.
    Shima-o-konohazuku. (Seven Island Scops-Owl.)

244 *Otus japonicus japonicus* Temminck and Schlegel.
In Siebold's Fauna Japon., Aves, p. 27, pl. 9, 1850.
    Konohazuku. (Japanese Scops-Owl.) B.

118 Ninox Hodgson (1837)

245 *Ninox scutulata scutulata* (Raffles).
Trans. Linn. Soc., XIII, p. 280, 1822.
     Awobazuku. (Brown Owlet.) B.

119 Ægolius Kaup (1829)

246 *Ægolius funereus sibiricus* (Buturlin).
Nasha Okhota, Juni-Nummer 1910, p. 78.
     Kimme-fukuro. (Siberian Teng-
malm's Owl.) (R.v.)

244 *Ægolius funereus funereus* (L.).
Syst. Nat., ed. x, p. 93, 1758.
     Tengmalm's Owl. (R.v.)

120 Glaucidium Boie (1826)

247 *Glaucidium passerinum orientale* Tac-
zanowski.
Faune Orn. Sib. Orient., I, p. 128, 1891.
     Suzume-fukuro. (Oriental Pigmy
Owl.) (R.v.)

121 Bubo Duméril (1806)

248 *Bubo bubo tenuipes* Clark.
Proc. U.S. Nat. Mus., XXXII, p. 470, 1907.
     Washi-mimizuku. (Clark's Eagle-
Owl.)

245 *Bubo bubo bubo* (L.).
Syst. Nat., ed. x, p. 92, 1758.
     Eagle-Owl. (R.v.)

249 *Bubo blakistoni blakistoni* Seebohm.
Proc. Zool. Soc. London, 1883, p. 466.
     Shima-fukuro. (Blakiston's Eagle-
Owl.) B.

122 Athene Boie (1822)

246 *Athene noctua vidalii* (A. E. Brehm).
Allg. D. Naturh. Zeitung, 1857, p. 440.
     Little Owl. B.

Order ACCIPITRES

Family AQUILIDÆ

123 Gyps Savigny (1809)

247 *Gyps fulvus fulvus* (Hablizl).
Neue Nordische Beytraege, IV, p. 58, 1783.
     Griffon-Vulture. (Exc.v.)

124 Neophron Savigny (1809)

248 *Neophron percnopterus percnopterus* (L.).
Syst. Nat., ed. x, p. 87, 1758.
     Egyptian Vulture. (Exc.v.)

125 Circus Lacépède (1806)

250 *Circus æruginosus æruginosus* (L.).
Syst. Nat., ed. X, p. 91, 1758.
Chūhi. (Marsh-Harrier.)

249 *Circus æruginosus æruginosus* (L.).
Syst. Nat., ed. X, p. 91, 1758.
Marsh-Harrier. B.

250 *Circus pygargus* (L.).
Syst. Nat., ed. X, p. 89, 1758.
Montagu's Harrier.

251 *Circus spilonotus* Kaup.
Jardine's Contr. Orn. for 1850, p. 59.
Shiberia-chuhi. (Eastern Marsh-Harrier.)

252 *Circus cyaneus cyaneus* (L.).
Syst. Nat., ed. XII, p. 126, 1766.
Haiiro-chuhi. (Hen-Harrier.)

251 *Circus cyaneus cyaneus* (L.).
Syst. Nat., ed. XII, p. 126, 1766.
Hen-Harrier.

126 Buteo Lacépède (1799)

253 *Buteo buteo japonicus* (Temminck and Schlegel).
In Siebold's Fauna Japon., Aves, p. 16, pl. VI and VI *b* (Text 1844, pl. 1845).
Nosuri. (Japanese Buzzard.) B.

252 *Buteo buteo buteo* (L.).
Syst. Nat., ed. X, p. 90, 1758.
Common Buzzard. B.

253 *Buteo buteo vulpinus* Gloger.
Abänd. Vög., Klima, p. 141, 1833.
Steppe-Buzzard. (Exc.v.)

254 *Buteo ferox hemilasius* Temminck and Schlegel.
In Siebold's Fauna Japon., Aves, p. 16, pl. 7 (Text 1844, pl. 1845).
O-nosuri. (Upland Buzzard.)

255 *Buteo lagopus pallidus* (Menzbier).
Orn. Turkestan, I, p. 163, 1888.
Keashi-nosuri. (Siberian Rough-legged Buzzard.) B.

254 *Buteo lagopus lagopus* (Brünnich).
Orn. Bor., p. 4, 1764.
Rough-legged Buzzard.

127 Aquila Brisson (1760)

256 *Aquila chrysaëtus chrysaëtus* (L.).
Syst. Nat., ed. X, p. 88, 1758.
Inu-washi. (Golden Eagle.) B.

255 *Aquila chrysaëtus chrysaëtus* (L.).
Syst. Nat., ed. X, p. 88, 1758.
Golden Eagle. B.

257 *Aquila heliaca heliaca* Savigny.
Descr. Égypte, Syst. Ois., p. 82, pl. 12, 1809.
Katashiro-washi. (Imperial Eagle.)

258 *Aquila clanga* Pallas.
Zoogr. Rosso-Asiat., I, p. 351, 1827.
Karafuto-washi. (Spotted Eagle.)

256 *Aquila clanga* Pallas.
Zoogr. Rosso-Asiat., I, p. 351, 1827.
Spotted Eagle. (R.v.)

128 Spizaëtus Vieillot (1816)

259 *Spizaëtus nipalensis orientalis* Temminck and Schlegel.
In Siebold's Fauna Japon., Aves, p. 7, pl. III (Text 1844, pl. 1845).
Kumataka. (Japanese Hawk-Eagle.) B.

129 Butastur Hodgson (1843)

260 *Butastur indicus* (Gmelin).
  Syst. Nat., I, 1, p. 264, 1788.
    Sashiba. (Eastern Buzzard-Hawk.)
    B.

130 Haliæetus Savigny (1809)

261 *Haliæetus albicilla* (L.).
  Syst. Nat., ed. x, p. 89, 1758.
    Ojiro-washi. (White-tailed Eagle.)
    B.

257 *Haliæetus albicilla* (L.).
  Syst. Nat., ed. x, p. 89, 1758.
    White-tailed Eagle. B.

262 *Haliæetus pelagicus* (Pallas).
  Zoogr. Rosso-Asiat., I, p. 343 and pl., 1827.
    O-washi. (Steller's Sea-Eagle.) B.

131 Accipiter Brisson (1760)

258 *Accipiter gentilis gentilis* (L.).
  Syst. Nat., ed. x, p. 89, 1758.
    Goshawk. B.

263 *Accipiter gentilis fujiyamæ* Swan and
  Hartert.
  Bull. B. O. C. XLIII, p. 170, June, 1923.
    O-taka. (Siberian Goshawk.) B.

259 *Accipiter gentilis atricapillus* (Wilson).
  Amer. Orn., VI, p. 80, pl. 52, fig. 3, 1812.
    American Goshawk. (Exc.v.)

264 *Accipiter nisus nisosimilis* (Tickell).
  Journ. As. Soc. Bengal, II, p. 571, 1833.
    Konori ♂ Haitaka ♀. (Jungle Spar-
    row-Hawk.) B.

260 *Accipiter nisus nisus* (L.).
  Syst. Nat., ed. x, p. 92, 1758.
    Sparrow-Hawk. B.

265 *Accipiter nisus pallens* Stejneger.
  Proc. U.S. Nat. Mus., XVI, p. 625, 1893.
    Haiiro-haitaka. (Kamschatkan
    Sparrow-Hawk.)

266 *Accipiter gularis* (Temminck and Schlegel).
  Siebold's Fauna Japon., Aves, p. 5, pl. 2 (Text
  1844, pl. 1845).
    Essai ♂ Tsumi ♀. (Japanese Spar-
    row-Hawk.) B.

132 Milvus Lacépède (1799)

261 *Milvus milvus milvus* (L.).
  Syst. Nat., ed. x, p. 89, 1758.
    Kite. B.

262 *Milvus migrans migrans* (Boddaert).
  Tables Planches Enl., No. 472, p. 28, 1783.
    Black Kite. (Exc.v.)

267 *Milvus lineatus* (Gray).
  Hardwicke's Ill. Ind. Zool., I, p. 1, pl. 18, 1832.
    Tobi. (Black-eared Kite.) B.

133 Pernis Cuvier (1817)

268 *Pernis apivorus orientalis* Taczanowski.
Faune Orn. Sib. Or., I, p. 50, 1891.
Hachikuma. (Siberian Honey-Buzzard.) B.

263 *Pernis apivorus apivorus* (L.).
Syst. Nat., ed. X, p. 91, 1758.
Honey-Buzzard.

134 Pandion Savigny (1809)

269 *Pandion haliætus haliætus* (L.).
Syst. Nat., ed. X, p. 91, 1758.
Misago. (Osprey.) B.

264 *Pandion haliætus haliætus* (L.).
Syst. Nat., ed. X, p. 91, 1758.
Osprey. B.

Family FALCONIDÆ

135 Falco Linnæus (1758)

265 *Falco rusticolus rusticolus* L.
Syst. Nat., ed. X, p. 88, 1758.
Gyr-Falcon. (Exc.v.)

266 *Falco rusticolus islandus* Brünnich.
Orn. Bor., p. 2, 1764.
Iceland Falcon. (R.v.)

270 *Falco rusticolus candicans* Gmelin.
Syst. Nat., I, p. 275, 1788.
Shiro-hayabusa. (Greenland Falcon).

267 *Falco rusticolus candicans* Gmelin.
Syst. Nat., I, I, p. 275, 1788.
Greenland Falcon.

271 *Falco peregrinus calidus* Latham.
Ind. Orn., I, p. 41, 1790. Ex Latham, Suppl. Gen. Synops., p. 35, no. 112.
Hayabusa. (Siberian Peregrine Falcon.)

268 *Falco peregrinus peregrinus* Tunstall.
Ornithologia Britannica, p. 1, 1771.
Peregrine Falcon. B.

272 *Falco peregrinus pealei* Ridgway.
Bull. Essex Inst., V, p. 201, 1873.
O-hayabusa. (Peale's Peregrine Falcon.)

269 *Falco peregrinus anatum* Bonaparte.
Geogr. and Comp. List, p. 4, 1838.
North American Peregrine. (Exc.v.)

273 *Falco subbuteo jakutensis* (Buturlin).
Nascha Ochota, IV, 6, p. 71, 1910.
Chigo-hayabusa. (Kamschatkan Hobby.)

270 *Falco subbuteo subbuteo* L.
Syst. Nat., ed. X, p. 89, 1758.
Hobby. B.

274 *Falco columbarius insignis* (Clark).
Proc. U.S. Nat. Mus., XXXII, p. 470, 1907.
Ko-chogenbo. (Asiatic Merlin.)

271 *Falco columbarius æsalon* Tunstall.
Orn. Brit., p. 1, 1771.
Merlin. B.

272 *Falco vespertinus vespertinus* L.
Syst. Nat., ed. XII, 1, p. 129, 1766.
Red-footed Falcon. (R.v.)

275 *Falco tinnunculus japonicus* Temminck and Schlegel.
Siebold's Fauna Japonica, Aves, p. 2; pl. 1 and 1 *b* (Text 1844, pl. 1845).
Chogenbo. (Japanese Kestrel.) B.

273 *Falco tinnunculus tinnunculus* L.
Syst. Nat., ed. X, p. 90, 1758.
Kestrel. B.

274 *Falco naumanni naumanni* Fleischer.
Sylvan, Jahrbuch auf 1817 und 1818, p. 174, 1818.
Lesser Kestrel. (Exc.v.)

Family PHAËTHONTIDÆ

136   Scæophæthon Mathews (1913)

276 *Scæophæthon rubricauda brevirostris*
Mathews.
    Birds of Austral., vol. IV, p. 303, June 1915.
       Akao-nettaicho. (Bonin Island
       Ruddy Tropic Bird.) (Exc.v.)

137   Leptophæthon Mathews (1913)

277 *Leptophæthon lepturus dorotheae* (Mathews).
    Austral. Av. Rec., vol. II, pt. I, p. 7, Aug. 1913.
       Shirao-nettaicho. (Pacific White-
       tailed Tropic Bird.) (Exc.v.)

Family SULIDÆ

138   Sula Brisson (1760)

275 *Sula bassana* (L.).
    Syst. Nat., ed. X, p. 133, 1758.
       Gannet. B.

278 *Sula leucogaster plotus* (Forster).
    Descr. Anim., ed. Licht., p. 278, 1844.
       Katsuwodori. (New Caledonian
       Brown Gannet.)

Order STEGANOPODES

Family PHALACROCORACIDÆ

139   Phalacrocorax Brisson (1760)

279 *Phalacrocorax carbo* [(?)*sinensis*] (Shaw
and Nodder).
    Nat. Misc., XIII, pl. 529, Text, 1801.
       Umi-u. (Chinese Cormorant.) B.

276 *Phalacrocorax carbo carbo* (L.).
    Syst. Nat., ed. X, p. 133, 1758.
       Cormorant. B.

280 *Phalacrocorax capillatus* (Temminck
and Schlegel).
    Siebold's Fauna Japon., Aves, pl. 83, 83 B, 1850.
       Kawa-u. (Temminck's Cormorant.)
       B.

277 *Phalacrocorax aristotelis aristotelis* (L.).
    Faun. Svec., Faunula, p. 5, no. 146, ex p. 51, 1761.
       Shag. B.

281 *Phalacrocorax urile* (Gmelin).
    Syst. Nat., I, II, p. 575, 1789.
       Chishima-ugarasu. (Red-faced Shag.)

282 *Phalacrocorax pelagicus pelagicus* Pallas.
    Zoogr. Rosso-Asiat., II, p. 303, 1827.
       Ugarasu or Hime-u. (Pelagic Shag.)
       B.

Family FREGATIDÆ

140 Fregata Brisson (1760)

283 *Fregata ariel* [(?)*ariel*] (Gould).
In Gray's Gen. B., III, p. 669, 1845.
   Gunkandori. (Lesser Frigate Bird.)
   (R.v.)

Family PELECANIDÆ

141 Pelecanus Linnæus (1758)

284 *Pelecanus crispus* Bruch.
Isis, 1832, p. 1109.
   Garancho. (Dalmatian Pelican.)
   (R.v.)

Order ANSERES

Family ANATIDÆ

142 Cygnus Bechstein (1803)

285 *Cygnus cygnus* (L.).
Syst. Nat., ed. X, p. 122, 1758.
   O-hakucho. (Whooper Swan.)

278 *Cygnus cygnus* (L.).
Syst. Nat., ed. X, p. 122, 1758.
   Whooper Swan. B.

286 *Cygnus bewickii jankowskii* Alphéraky.
Priroda i Okhota (Nature and Sport), Sept.
1904, p. 10.
   Hakucho. (Eastern Bewick's Swan.)

279 *Cygnus bewickii bewickii* Yarrell.
Trans. Linn. Soc. London, XVI, II, p. 453,
1830.
   Bewick's Swan.

280 *Cygnus olor* (Gmelin).
Syst. Nat., I, II, p. 501, 1789.
   Mute Swan. B.

143 Cygnopsis Brandt (1836)

287 *Cygnopsis cygnoides* (L.).
Syst. Nat., ed. X, p. 122, 1758.
   Sakatsura-gan. (Chinese Goose.) B.

144 Anser Brisson (1760)

288 *Anser anser* (L.).
Syst. Nat., ed. X, p. 123, 1758.
   Haiiro-gan. (Grey Lag-Goose.)
   (Exc.v.)

281 *Anser anser* (L.).
Syst. Nat., ed. X, p. 123, 1758.
   Grey Lag-Goose. B.

289 *Anser fabalis fabalis* (Latham).
Gen. Syn., Suppl., I, p. 297, 1787.
   Hime-hishikui. (Bean-Goose.)

282 *Anser fabalis fabalis* (Latham).
Gen. Syn., Suppl., I, p. 297, 1787.
   Bean-Goose.

290 *Anser fabalis sibiricus* (Alphéraky).
Geese Europe and Asia, p. 140, pls. 10 and 23,
1905.
   O-hishikui. (Middendorff's Goose.)

291 *Anser fabalis mentalis* Oates.
Man. Game-B. India, II, p. 77, 1899.
   Hashibuto-hishikui.   (Thick-billed
Goose.)

292 *Anser fabalis serrirostris* Swinhoe.
Proc. Zool. Soc. London, 1871, p. 417.
   Hishikui. (Eastern Bean-Goose.)

293 *Anser albifrons gambelli* Hartlaub.
Rev. and Mag. Zool., 1852, p. 7.
   O-magan. (North American White-
fronted Goose.)

294 *Anser albifrons albifrons* (Scopoli).
Annus I Historico-Natur., p. 69, 1769.
   Magan. (White-fronted Goose.) B.

283 *Anser albifrons albifrons* (Scopoli).
Annus I Historico-Natur., p. 69, 1769.
   White-fronted Goose.

295 *Anser erythropus* (L.).
Syst. Nat., ed. X, p. 123, 1758.
   Karigane.   (Lesser White-fronted
Goose.)

284 *Anser erythropus* (L.).
Syst. Nat., ed. X, p. 123, 1758.
   Lesser White-fronted Goose. (Exc.v.)

285 *Anser brachyrhynchus* Baillon.
Mém. Soc. R. Abbeville, 1833, p. 74.
   Pink-footed Goose.

296 *Anser hyperboreus hyperboreus* Pallas.
Spicilegia Zool., fasc. VI, p. 25, 1769.
   Haku-gan. (Snow-Goose.)

286 *Anser hyperboreus hyperboreus* Pallas.
Spicilegia Zool., fasc. VI, p. 25, 1769.
   Snow-Goose.

287 *Anser hyperboreus nivalis* (Forster).
Philos. Trans., LXII, p. 433, 1772.
   Greater Snow-Goose. (Exc.v.)

145 Branta Scopoli (1769)

297 *Branta bernicla nigricans* (Lawrence).
Ann. Lyc. Nat. Hist., New York, IV, p. 171,
pl. XII, 1846.
   Koku-gan. (Black Brent Goose.)

288 *Branta bernicla bernicla* (L.).
Syst. Nat., ed. X, p. 124, 1758.
   Brent Goose.

289 *Branta bernicla glaucogaster* C. L.
Brehm.
Isis, 1830, p. 996.
   American Brent Goose.

290 *Branta leucopsis* (Bechstein).
Orn. Taschenb., II, p. 424, 1803.
   Barnacle-Goose.

291 *Branta ruficollis* (Pallas).
Spicilegia Zool., fasc. VI, p. 21, pl. IV, 1769.
   Red-breasted Goose. (Exc.v.)

298 *Branta canadensis hutchinsii* (Richard-
son).
In Swainson and Richardson, Fauna Bor.-
Amer., II, Birds, p. 470, 1831.
   Shijukara-gan. (Hutchin's Goose.) B.

146 Tadorna Fleming (1822)

299 *Tadorna tadorna* (L.).
Syst. Nat., ed. x, p. 122, 1758.
Tsukushi-gamo. (Sheld-Duck.)

292 *Tadorna tadorna* (L.).
Syst. Nat., ed. x, p. 122, 1758.
Sheld-Duck. B.

147 Casarca Bonaparte (1838)

300 *Casarca ferruginea* (Pallas).
Vroeg's Cat., Adumbratiuncula, p. 5, 1764.
Aka-tsukushigamo. (Ruddy Sheld-Duck.)

293 *Casarca ferruginea* (Pallas).
Vroeg's Cat., Adumbratiuncula, p. 5, 1764.
Ruddy Sheld-Duck. (R.v.)

148 Anas Linnæus (1758)

301 *Anas platyrhyncha platyrhyncha* L.
Syst. Nat., ed. x, p. 125, 1758.
Magamo. (Wild Duck or Mallard.) B.

294 *Anas platyrhyncha platyrhyncha* L.
Syst. Nat., ed. x, p. 125, 1758.
Wild Duck or Mallard. B.

302 *Anas poecilorhyncha zonorhyncha* Swinhoe.
Ibis, 1866, p. 394.
Karu-gamo. (Dusky Mallard.) B.

149 Chaulelasmus Bonaparte (1838)

303 *Chaulelasmus streperus* L.
Syst. Nat., ed. x, p. 125, 1758.
Okayoshi-gamo. (Gadwall.)

295 *Chaulelasmus streperus* L.
Syst. Nat., ed. x, p. 125, 1758.
Gadwall. B.

150 Nettion Kaup (1829)

304 *Nettion formosum* (Georgi).
Bemerk. Reise Russ. Reich., I, p. 168, 1775.
Tomoe-gamo. (Spectacled Teal.)

305 *Nettion crecca crecca* (L.).
Syst. Nat., ed. x, p. 126, 1758.
Kogamo. (Teal.) B.

296 *Nettion crecca crecca* (L.).
Syst. Nat., ed. x, p. 126, 1758.
Teal. B.

306 *Nettion crecca carolinensis* (Gmelin).
Syst. Nat., I, II, p. 533, 1789.
America-kogamo. (American Green-winged Teal.) (Exc.v.)

297 *Nettion crecca carolinensis* (Gmelin).
Syst. Nat., I, II, p. 533, 1789.
American Green-winged Teal. (Exc. v.)

151 Eunetta Bonaparte (1856)

307 *Eunetta falcata* (Georgi).
Bemerk. Reise Russ. Reich., I, p. 167, 1775.
Yoshi-gamo. (Falcated Teal.)

152 Mareca Stephens (1824)

308 *Mareca penelope* (L.).
Syst. Nat., ed. x, p. 126, 1758.
Hidori-gamo. (Wigeon.) B.

298 *Mareca penelope* (L.).
Syst. Nat., ed. x, p. 126, 1758.
Wigeon. B.

309 *Mareca americana* (Gmelin).
Syst. Nat., I, II, p. 526, 1789.
America-hidori. (American Wigeon.)

299 *Mareca americana* (Gmelin).
Syst. Nat., I, II, p. 526, 1789.
American Wigeon. (R.v.)

153 Dafila Stephens (1824)

310 *Dafila acuta acuta* (L.).
Syst. Nat., ed. x, p. 126, 1758.
Onaga-gamo. (Pintail.)

300 *Dafila acuta acuta* (L.).
Syst. Nat., ed. x, p. 126, 1758.
Pintail. B.

154 Querquedula Stephens (1824)

311 *Querquedula querquedula* (L.).
Syst. Nat., ed. x, p. 126, 1758.
Shimaaji. (Garganey.)

301 *Querquedula querquedula* (L.).
Syst. Nat., ed. x, p. 126, 1758.
Garganey. B.

302 *Querquedula discors* (L.).
Syst. Nat., ed. xii, p. 205, 1766.
American Blue-winged Teal. (Exc.v.)

155 Aix Boie (1828)

312 *Aix galericulata* (L.).
Syst. Nat., ed. x, p. 128, 1758.
Oshidori. (Mandarin Duck.) B.

156 Spatula Boie (1822)

313 *Spatula clypeata* (L.).
Syst. Nat., ed. x, p. 124, 1758.
Hashibiro-gamo. (Shoveler.) B.

303 *Spatula clypeata* (L.).
Syst. Nat., ed. x, p. 124, 1758.
Shoveler. B.

157 Netta Kaup (1829)

304 *Netta rufina* (Pallas).
Reise d. versch. Prov. d. Russ. Reichs, ii,
p. 713, 1773.
Red-crested Pochard. (R.v.)

158 Nyroca Fleming (1822)

314 *Nyroca nyroca baeri* (Radde).
Reise S. v. O. Sibirien, ii, p. 376, pl. 15, 1863.
Aka-hajiro. (Siberian White-eyed
Pochard.)

305 *Nyroca nyroca nyroca* (Güldenstädt).
Nov. Comm. Sc. Petrop., xiv, i, p. 403, 1770.
White-eyed Pochard. (R.v.)

315 *Nyroca ferina ferina* (L.).
Syst. Nat., ed. x, p. 126, 1758.
Hoshi-hajiro. (Pochard.)

306 *Nyroca ferina ferina* (L.).
Syst. Nat., ed. x, p. 126, 1758.
Pochard. B.

316 *Nyroca marila mariloides* (Vigors).
Zool. Beechey's Voy. Blossom, p. 31, 1839.
Suzu-gamo. (Eastern Scaup-Duck.)

307 *Nyroca marila marila* (L.).
Fauna Svecica, ed. ii, p. 39, 1761.
Scaup-Duck. B.

317 *Nyroca fuligula* (L.).
Syst. Nat., ed. x, p. 128, 1758.
Kinkuro-hajiro. (Tufted Duck.)

308 *Nyroca fuligula* (L.).
Syst. Nat., ed. x, p. 128, 1758.
Tufted Duck. B.

159 Bucephala Baird (1860)

318 *Bucephala clangula clangula* (L.).
Syst. Nat., ed. x, p. 125, 1758.
Hojiro-gamo. (Golden-eye.)

309 *Bucephala clangula clangula* (L.).
Syst. Nat., ed. x, p. 125, 1758.
Golden-eye.

310 *Bucephala albeola* (L.).
Syst. Nat., ed. x, p. 124, 1758.
Buffel-headed Duck. (Exc.v.)

160 Clangula Leach (1819)

319 *Clangula hyemalis* (L.).
Syst. Nat., ed. x, p. 126, 1758.
Kori-gamo. (Long-tailed Duck.)

311 *Clangula hyemalis* (L.).
Syst. Nat., ed. x, p. 126, 1758.
Long-tailed Duck. B.

161 Histrionicus Lesson (1828)

320 *Histrionicus histrionicus pacificus*
Brooks.
Bull. Mus. Comp. Zool. Harvard College, LIX,
p. 393, 1915.
Shinori-gamo. (Pacific Harlequin-
Duck.) B.

312 *Histrionicus histrionicus histrionicus*
(L.).
Syst. Nat., ed. x, p. 127, 1758.
Harlequin-Duck. (Exc.v.)

162 Melanitta Boie (1822)

321 *Melanitta nigra americana* Swainson.
Swainson and Richardson's Fauna Bor. Amer.,
II, p. 450, 1832.
Kuro-gamo. (American Black
Scoter.)

313 *Melanitta nigra nigra* (L.).
Syst. Nat., ed. x, p. 123, 1758.
Common Scoter. B.

322 *Melanitta fusca stejnegeri* Ridgway.
Man. N. Amer. B., p. 112, 1887.
Birodo-kinkuro. (Eastern Velvet-
Scoter.)

314 *Melanitta fusca fusca* (L.).
Syst. Nat., ed. x, p. 123, 1758.
Velvet-Scoter.

315 *Melanitta perspicillata* (L.).
Syst. Nat., ed. x, p. 125, 1758.
Surf-Scoter. (R.v.)

163 Polysticta Eyton (1836)

323 *Polysticta stelleri* (Pallas).
Spicil. Zool., fasc. VI, p. 35, 1769.
Ko-kewatagamo. (Steller's Eider.)

316 *Polysticta stelleri* (Pallas).
Spicil. Zool., fasc. VI, p. 35, 1769.
Steller's Eider. (Exc.v.)

164 Somateria Leach (1819)

317 *Somateria mollissima mollissima* (L.).
Syst. Nat., ed. x, p. 124, 1758.
Common Eider. B.

324 *Somateria spectabilis* (L.).
Syst. Nat., ed. x, p. 123, 1758.
Kewata-gamo. (King-Eider.)

318 *Somateria spectabilis* (L.).
Syst. Nat., ed. x, p. 123, 1758.
King-Eider. (R.v.)

165 Mergus Linnæus (1758)

325 *Mergus merganser merganser* L.
Syst. Nat., ed. x, p. 129, 1758.
Kawa-aisa. (Goosander.)

319 *Mergus merganser merganser* L.
Syst. Nat., ed. x, p. 129, 1758.
Goosander. B.

326 *Mergus merganser orientalis* Gould.
Proc. Zool. Soc., London, 1845, p. 1.
Ko-kawaaisa. (Eastern Goosander.)

H.

327 *Mergus serrator* L.
  Syst. Nat., ed. x, p. 129, 1758.
    Umi-aisa. (Red-breasted Mer-
    ganser.) B.

328 *Mergus albellus* L.
  Syst. Nat., ed. x, p. 129, 1758.
    Miko-aisa. (Smew.)

320 *Mergus serrator* L.
  Syst. Nat., ed. x, p. 129, 1758.
    Red-breasted Merganser. B.

321 *Mergus cucullatus* L.
  Syst. Nat., ed. x, p. 129, 1758.
    Hooded Merganser. (R.v.)

322 *Mergus albellus* L.
  Syst. Nat., ed. x, p. 129, 1758.
    Smew.

Order PHŒNICOPTERI

Family PHŒNICOPTERIDÆ

166 Phœnicopterus Linnæus (1758)

323 *Phœnicopterus ruber antiquorum* Tem-
    minck.
  Man. d'Orn., ed. II, p. 587, 1820.
    Flamingo. (Exc.v.)

Order GRESSORES

Family CICONIIDÆ

167 Ciconia Brisson (1760)

329 *Ciconia ciconia boyciana* Swinhoe.
  Proc. Zool. Soc. London, 1873, p. 513.
    Konotori. (Japanese Stork.) B.

330 *Ciconia nigra* (L.).
  Syst. Nat., ed. x, p. 142, 1758.
    Nabe-ko. (Black Stork.) (R.v.)

324 *Ciconia ciconia ciconia* (L.).
  Syst. Nat., ed. x, p. 142, 1758.
    White Stork.

325 *Ciconia nigra* (L.).
  Syst. Nat., ed. x, p. 142, 1758.
    Black Stork. (R.v.)

Family IBIDIDÆ

168 Platalea Linnæus (1758)

331 *Platalea leucorodia major* Temminck
    and Schlegel.
  Siebold's Fauna Japon., Aves, p. 119, p. 75, 1849.
    Herasagi. (Japanese Spoonbill.)
    (R.v.)

332 *Platalea minor* Temminck and Schle-
    gel.
  Siebold's Fauna Japon., Aves, p. 120, pl. 76,
  1849.
    Kurotsura-herasagi. (Black-faced
    Spoonbill.) (R.v.)

326 *Platalea leucorodia leucorodia* L.
  Syst. Nat., ed. x, p. 139, 1758.
    Spoonbill.

169 Plegadis Kaup (1829)

327 *Plegadis falcinellus falcinellus* (L.).
  Syst. Nat., ed. XII, p. 241, 1766.
    Glossy Ibis.

170 Nipponia Reichenbach (1852)

333 *Nipponia nippon* (Temminck).
Pl. Color., 551, 1835.
  Toki. (Japanese Crested Ibis.) (R.v.)

171 Threskiornis Gray (1842)

334 *Threskiornis melanocephalus* (Latham).
Ind. Orn., II, p. 709, 1790.
  Kurotoki. (White Ibis.) (R.v.)

Family ARDEIDÆ

172 Ardea Linnæus (1758)

335 *Ardea cinerea jouyi* Clark.
Proc. U.S. Nat. Mus., XXXII, p. 468, 1907.
  Awosagi. (Eastern Common Heron.)
  B.

328 *Ardea cinerea cinerea* L.
Syst. Nat., ed. X, p. 143, 1758.
  Common Heron. B.

336 *Ardea purpurea manillensis* Meyen.
Acta Acad. Leop. Carol., XVI, Suppl. p. 102;
id., Reise um die Erde, III, p. 226, 1831.
  Murasaki-sagi. (Eastern Purple
  Heron.) (R.v.)

329 *Ardea purpurea purpurea* L.
Syst. Nat., ed. XII, p. 236, 1766.
  Purple Heron. (R.v.)

173 Egretta Forster (1817)

337 *Egretta alba alba* (L.).
Syst. Nat., ed. X, p. 144, 1758.
  Dai-sagi. (Great White Heron.)

330 *Egretta alba alba* (L.).
Syst. Nat., ed. X, p. 144, 1758.
  Great White Heron. (R.v.)

338 *Egretta alba modesta* (Gray).
Zool. Misc., p. 19, 1831.
  Komomojiro. (Eastern Great White
  Heron.) B.

339 *Egretta intermedia intermedia* (Wag-
ler).
Isis, 1829, p. 659.
  Chu-sagi. (Plumed Egret.) B.

340 *Egretta garzetta garzetta* (L.).
Syst. Nat., ed. XII, p. 237, 1766.
  Ko-sagi. (Little Egret.) B.

331 *Egretta garzetta garzetta* (L.).
Syst. Nat., ed. XII, p. 237, 1766.
  Little Egret. (R.v.)

174 Demiegretta Blyth (1846)

341 *Demiegretta sacra ringeri* Stejneger.
Proc. U.S. Nat. Mus., X, p. 300, 1887.
  Kuro-sagi. (Japanese Reef Heron.)

175 Bubulcus Bonaparte (1855)

342 *Bubulcus ibis coromandus* (Boddaert).
Tables Planches Enl., No. 472, p. 28, 1783.
  Amasagi. (Indian Cattle Egret.) B.

332 *Bubulcus ibis ibis* (L.).
Syst. Nat., ed. X, p. 144, 1758.
  Buff-backed Heron. (Exc.v.)

176 Ardeola Boie (1822)

343 *Ardeola bacchus* (Bonaparte).
Consp. Gen. Av., II, p. 127, 1855.
  Akagashira-sagi. (Chinese Pond
Heron.) (R.v.)

333 *Ardeola ralloides* (Scopoli).
Annus I Historico-Natur., p. 88, 1769.
  Squacco Heron. (R.v.)

177 Butorides Blyth (1852)

344 *Butorides striatus amurensis* Schrenck.
Reis. Amur-Lande, I, II, p. 441, 1860.
  Sasagoi. (Amur Green Heron.) B.

178 Nycticorax Forster (1817)

345 *Nycticorax nycticorax nycticorax* (L.).
Syst. Nat., ed. X, p. 142, 1758.
  Goi-sagi. (Night-Heron.) B.

334 *Nycticorax nycticorax nycticorax* (L.).
Syst. Nat., ed. X, p. 142, 1758.
  Night-Heron.

179 Gorsachius Bonaparte (1855)

346 *Gorsachius goisagi* (Temminck).
Pl. Color., 582, 1835.
  Mizo-goi. (Japanese Night-Heron.)
B.

180 Ixobrychus Billberg (1828)

335 *Ixobrychus minutus minutus* (L.).
Syst. Nat., ed. XII, p. 240, 1766.
  Little Bittern.

347 *Ixobrychus sinensis* (Gmelin).
Syst. Nat., I, II, p. 642, 1789.
  Yoshi-goi. (Chinese Little Bittern.)
B.

181 Nannocnus Stejneger (1887)

348 *Nannocnus eurhythmus* (Swinhoe).
Ibis, 1873, p. 74, pl. 2.
  O-yoshigoi. (Schrenck's Little Bittern.) B.

182 Botaurus Stephens (1819)

349 *Botaurus stellaris stellaris* (L.).
Syst. Nat., ed. X, p. 144, 1758.
  Sankano-goi. (Bittern.) B.

336 *Botaurus stellaris stellaris* (L.).
Syst. Nat., ed. X, p. 144, 1758.
  Bittern. B.

337 *Botaurus lentiginosus* (Montagu).
Orn. Dict., Suppl. (under Freckled Heron),
text and plate, 1813.
  American Bittern. (R.v.)

Order ALECTORIDES

Family OTIDIDÆ

183 Otis Linnæus (1758)

350 *Otis tarda dybowskii* Taczanowski.
Journ. f. Orn., 1874, p. 331.
  No-gan. (Siberian Bustard.) (R.v.)

338 *Otis tarda tarda* L.
Syst. Nat., ed. x, p. 154, 1758.
  Great Bustard. (R.v.)

339 *Otis tetrax orientalis* Hartert.
Nov. Zool., 1916, p. 339.
  Eastern Little Bustard.

184 Chlamydotis Lesson (1839)

340 *Chlamydotis undulata macqueenii* (Gray and Hardwicke).
Illustr. Ind. Zool., II, pl. 47, 1834.
  Macqueen's Bustard. (Exc.v.)

Family BALEARICIDÆ

185 Megalornis Gray (1841)

351 *Megalornis grus lilfordi* (Sharpe).
Cat. Birds, Brit. Mus., XXIII, p. 252, 1894.
  Kuro-zuru. (Eastern Common Crane.) (R.v.)

341 *Megalornis grus grus* (L.).
Syst. Nat., ed. x, p. 141, 1758.
  Common Crane. (R.v.)

352 *Megalornis leucogeranus* (Pallas).
Reise d. versch. Prov. Russ. Reichs, II, p. 714, 1773.
  Sodeguro-zuru. (Siberian White Crane.) (R.v.)

353 *Megalornis vipio* (Pallas).
Zoogr. Rosso-Asiat., II, p. 111, 1827.
  Mana-zuru. (White-naped Crane.) B.

354 *Megalornis japonensis* (P. L. S. Müller).
Natursystem, Suppl., p. 110, 1776.
  Tancho. (Japanese Crane.) (R.v.)

355 *Megalornis monachus* (Temminck).
Pl. Color., 555, 1835.
  Nabe-zuru. (Hooded Crane.)

186 Anthropoides Vieillot (1816)

356 *Anthropoides virgo* (L.).
Syst. Nat., ed. x, p. 141, 1758.
  Aneha-zuru. (Demoiselle Crane.) (R.v.)

Family RALLIDÆ

187 Rallus Linnæus (1758)

357 Rallus aquaticus indicus Blyth.
Journ. As. Soc. Bengal, XVIII, part 2, p. 820, 1849.
   Kuina. (Eastern Water-Rail.) B.

342 Rallus aquaticus aquaticus L.
Syst. Nat., ed. X, p. 153, 1758.
   Water-Rail. B.

188 Crex Bechstein (1803)

343 Crex crex (L.).
Syst. Nat., ed. X, p. 153, 1758.
   Corn Crake. B.

189 Porzana Vieillot (1816)

344 Porzana porzana (L.).
Syst. Nat., ed. XII, p. 262, 1766.
   Spotted Crake. B.

345 Porzana carolina (L.).
Syst. Nat., ed. X, p. 153, 1758.
   Carolina Crake. (Exc.v.)

358 Porzana pusilla pusilla (Pallas).
Reise d. versch. Prov. russ. Reichs, III, p. 700, 1776.
   Hime-kuina. (Pallas's Crake.) B.

346 Porzana pusilla intermedia (Hermann).
Obs. Zool., I, p. 198, 1804.
   Baillon's Crake.

347 Porzana parva (Scopoli).
Annus I Historico-Natur., p. 108, 1769.
   Little Crake.

359 Porzana noveboracensis exquisita Swinhoe.
Ann. and Mag. Nat. Hist. (4), XII, p. 376, 1873.
   Shima-kuina. (Swinhoe's Crake.)

190 Limnobænus Sundevall (1872)

360 Limnobænus fuscus erythrothorax (Temminck and Schlegel).
In Siebold's Fauna Japon., Aves, p. 121, pl. 78, 1849.
   Hi-kuina. (Japanese Ruddy Crake.) B.

191 Gallinula Brisson (1760)

361 Gallinula chloropus parvifrons Blyth.
Journ. As. Soc. Bengal, XII, p. 180, 1843.
   Ban. (Indian Moor-Hen.) B.

348 Gallinula chloropus chloropus (L.).
Syst. Nat., ed. X, p. 152, 1758.
   Moor-Hen. B.

192 Gallicrex Blyth (1849)

362 Gallicrex cinerea (Gmelin).
Syst. Nat., I, II, p. 702, 1789.
   Tsuru-kuina. (Water-Cock.) B.

193 Fulica Linnæus (1758)

363 Fulica atra atra L.
Syst. Nat., ed. X, p. 152, 1758.
   O-ban. (Coot.) B.

349 Fulica atra atra L.
Syst. Nat., ed. X, p. 152, 1758.
   Coot. B.

Order LIMICOLÆ

Family BURHINIDÆ

194 Burhinus Illiger (1811)

350 *Burhinus œdicnemus œdicnemus* (L.).
Syst. Nat., ed. X, p. 151, 1758.
Stone-Curlew. B.

Family CURSORIIDÆ

195 Cursorius Latham (1790)

351 *Cursorius cursor cursor* (Latham).
Gen. Synops. Birds, Suppl. 1, p. 293, 1787.
Cream-coloured Courser. (Exc.v.)

Family GLAREOLIDÆ

196 Glareola Brisson (1760)

364 *Glareola pratincola maldivarum*
Forster.
Faunula Indica, p. 11, 1795.
Tsubame-chidori. (Oriental Pratin-cole.) (Exc.v.)

352 *Glareola pratincola pratincola* (L.).
Syst. Nat., ed. XII, p. 345, 1766.
Collared Pratincole. (Exc.v.)

353 *Glareola nordmanni* Fischer.
Bull. Soc. Imp. Nat. Moscou, XV, p. 314, pl. 2, 1842.
Black-winged Pratincole. (Exc.v.)

Family CHARADRIIDÆ

197 Hæmatopus Linnæus (1758)

365 *Hæmatopus ostralegus osculans* Swin-hoe.
Proc. Zool. Soc. London, 1871, p. 405.
Miyakodori.   (Eastern Oyster-catcher.) B.

366 *Hæmatopus niger bachmani* Audubon.
B. America, IV, pl. 427, fig. 1, 1838.
Kuro-miyakodori. (West American Black Oyster-catcher.) (Exc.v.)

354 *Hæmatopus ostralegus ostralegus* L.
Syst. Nat., ed. X, p. 152, 1758.
Oyster-catcher. B.

198 Pluvialis Brisson (1760)

355 *Pluvialis apricarius apricarius* (L.).
Syst. Nat., ed. X, p. 150, 1758.
Southern Golden Plover. B.

356 *Pluvialis apricarius altifrons* (C. L. Brehm).
Handb. Naturg. Vög. Deutschl., p. 542, 1831.
Northern Golden Plover.

357 *Pluvialis dominicus dominicus* (P. L. S. Müller).
Natursystem, Suppl., p. 116, 1776.
American Golden Plover. (Exc.v.)

367 *Pluvialis dominicus fulvus* (Gmelin).
Syst. Nat., I, II, p. 687, 1789.
Munaguro. (Asiatic Golden Plover.)

358 *Pluvialis dominicus fulvus* (Gmelin).
Syst. Nat., I, II, p. 687, 1789.
Asiatic Golden Plover. (Exc.v.)

199 Squatarola Cuvier (1816)

368 *Squatarola squatarola squatarola* (L.).
Syst. Nat., ed. X, p. 149, 1758.
Daizen. (Grey Plover.)

359 *Squatarola squatarola squatarola* (L.).
Syst. Nat., ed. X, p. 149, 1758.
Grey Plover.

200 Eudromias Brehm (1831)

369 *Eudromias morinellus* (L.).
Syst. Nat., ed. X, p. 150, 1758.
Kobashi-chidori. (Dotterel.)

360 *Eudromias morinellus* (L.).
Syst. Nat., ed. X, p. 150, 1758.
Dotterel. B.

201 Cirrepidesmus Bonaparte (1856)

370 *Cirrepidesmus asiatica vereda* (Gould).
Proc. Zool. Soc. London, 1848, p. 38.
O-chidori. (Eastern Dotterel.)
(Exc.v.)

371 *Cirrepidesmus leschenaultii* (Lesson).
Dict. Sci. Nat. XLII (Levrault), p. 36, 1826.
O-medai-chidori. (Geoffroy's Sand-Plover.)

361 *Cirrepidesmus asiaticus asiaticus* (Pallas).
Reise d. versch. Prov. d. Russ. Reichs, II, p. 715, 1773.
Caspian Plover. (Exc.v.)

372 *Cirrepidesmus mongolus mongolus* (Pallas).
Reise d. versch. Prov. d. Russ. Reichs, III, p. 700, 1776.
Medai-chidori. (Mongolian Plover.)

202 Charadrius Linnæus (1758)

373 *Charadrius hiaticula tundræ* (Lowe).
Bull. Brit. Orn. Club, XXXVI, p. 7, 1915.
Hajiro-ko-chidori. (Lowe's Ringed Plover.) (Exc.v.)

362 *Charadrius hiaticula hiaticula* L.
Syst. Nat., ed. X, p. 150, 1758.
Ringed Plover. B.

363 *Charadrius semipalmatus* Bonaparte.
Journ. Acad. Nat. Sci., Philadelphia, V, p. 98, 1825.
Semi-palmated Ringed Plover. (Exc.v.)

374 *Charadrius alexandrinus alexandrinus* L.
Syst. Nat., ed. x, p. 150, 1758.
Hashiboso-shiro-chidori.   (Kentish Plover.) B.

375 *Charadrius alexandrinus dealbatus* (Swinhoe).
Proc. Zool. Soc. London, 1870, p. 138.
Shiro-chidori.   (Eastern Kentish Plover.) B.

376 *Charadrius dubius curonicus* Gmelin.
Syst. Nat., I, II, p. 692, 1789.
Ko-chidori. (Little Ringed Plover.) B.

377 *Charadrius dubius dubius* (Scopoli).
Del. Faunae et Florae Insubr., II, p. 93, 1786.
Minami-ko-chidori. (Southern Little Ringed Plover.)

378 *Charadrius placidus* Gray.
Cat. Mamm. Birds, etc. of Nepal and Tibet in Brit. Mus., ed. II, p. 70, 1863.
Ikaru-chidori. (Long-billed Ringed Plover.) B.

364 *Charadrius alexandrinus alexandrinus* L.
Syst. Nat., ed. x, p. 150, 1758.
Kentish Plover. B.

365 *Charadrius dubius curonicus* Gmelin.
Syst. Nat., I, II, p. 692, 1789.
Little Ringed Plover. (R.v.)

366 *Charadrius vociferus* L.
Syst. Nat., ed. x, p. 150, 1758.
Killdeer Plover. (Exc.v.)

203 Chettusia Bonaparte (1841)

367 *Chettusia gregaria* (Pallas).
Reise d. versch. Prov. d. Russ. Reichs, I, p. 456, 1771.
Sociable Plover. (Exc.v.)

204 Vanellus Brisson (1760)

379 *Vanellus vanellus* (L.).
Syst. Nat., ed. x, p. 148, 1758.
Ta-geri. (Lapwing.) B.

368 *Vanellus vanellus* (L.).
Syst. Nat., ed. x, p. 148, 1758.
Lapwing. B.

205 Microsarcops Sharpe (1896)

380 *Microsarcops cinereus* (Blyth).
Journ. As. Soc. Bengal, XI, p. 587, 1842.
Keri. (Grey-headed Wattled Lapwing.) B.

206 Arenaria Brisson (1760)

381 *Arenaria interpres interpres* (L.).
Syst. Nat., ed. x, p. 148, 1758.
Kyojo-shigi. (Turnstone.)

369 *Arenaria interpres interpres* (L.).
Syst. Nat., ed. x, p. 148, 1758.
Turnstone.

207 Bartramia Lesson (1831)

370 *Bartramia longicauda* (Bechstein).
Allg. Ueb. Vögel, IV, ii, p. 453, 1811.
Bartram's Sandpiper. (Exc.v.)

208 **Crocethia** Billberg (1828)

382 *Crocethia alba* (Pallas).
Vroeg's Cat. Coll., Adumbratiuncula, p. 7, 1764.
Miyubi-shigi. (Sanderling.)

371 *Crocethia alba* (Pallas).
Vroeg's Cat. Coll., Adumbratiuncula, p. 7, 1764.
Sanderling.

209 **Calidris** Anonymous (1804)

383 *Calidris canutus rogersi* (Mathews).
B. Australia, III, p. 270, pl. 163, 1913.
Ko-oba-shigi. (Eastern Knot.)

372 *Calidris canutus canutus* (L.).
Syst. Nat., ed. X, p. 149, 1758.
Knot.

384 *Calidris tenuirostris* (Horsfield).
Trans. Linn. Soc. London, XIII, p. 192, 1821.
Oba-shigi. (Great Knot.)

385 *Calidris ruficollis* (Pallas).
Reise d. versch. Prov. d. Russ. Reichs, III, p. 700, 1776.
Tonen. (Eastern Stint.)

386 *Calidris subminuta* (Middendorff).
Reise N. O. and O. Sibirien, II, 2, p. 222, pl. XIX, fig. 6, 1851.
Hibari-shigi. (Long-toed Stint.) B.

373 *Calidris minuta* (Leisler).
Nachträge zu Bechst. Naturg. Deutschl., p. 74, 1812.
Little Stint.

374 *Calidris minutilla* (Vieillot).
Nouv. Dict. d'Hist. Nat., nouv. éd., XXXIV, p. 466, 1819.
American Stint. (Exc.v.)

387 *Calidris temminckii* (Leisler).
Nachträge zu Bechst. Naturg. Deutschl., pp. 63–73, 1812.
Ojiro-tonen. (Temminck's Stint.) (R.v.)

375 *Calidris temminckii* (Leisler).
Nachträge zu Bechst. Naturg. Deutschl., pp. 63–73, 1812.
Temminck's Stint.

388 *Calidris acuminata* (Horsfield).
Trans. Linn. Soc. London, XIII, p. 192, 1821.
Uzura-shigi or Saru-shigi. (Siberian Pectoral Sandpiper.)

376 *Calidris acuminata* (Horsfield).
Trans. Linn. Soc. London, XIII, p. 192, 1821.
Siberian Pectoral Sandpiper. (Exc.v.)

389 *Calidris maculata* (Vieillot).
Nouv. Dict. d'Hist. Nat., nouv. éd., XXXIV, p. 465, 1819.
America-uzurashigi. (American Pectoral Sandpiper.) (Exc.v.)

377 *Calidris maculata* (Vieillot).
Nouv. Dict. d'Hist. Nat., nouv. éd., XXXIV, p. 465, 1819.
American Pectoral Sandpiper. (R.v.)

378 *Calidris bairdii* (Coues).
Proc. Acad. Sci. Philad., 1861, p. 194.
Baird's Sandpiper. (Exc.v.)

379 *Calidris fuscicollis* (Vieillot).
Nouv. Dict. d'Hist. Nat., nouv. éd., XXXIV, p. 461, 1819.
Bonaparte's Sandpiper. (Exc.v.)

390 *Calidris maritima couesi* (Ridgway).
Bull. Nutt. Orn. Club, V, p. 160, 1880.
Chishima-shigi. (Aleutian Sandpiper.)

380 *Calidris maritima maritima* (Brünnich).
Orn. Bor., p. 54, 1764.
Purple Sandpiper.

391 *Calidris alpina sakhalina* (Vieillot).
Nouv. Dict. d'Hist. Nat., III, p. 359, 1816.
Hama-shigi. (Pacific Dunlin.) B.

381 *Calidris alpina alpina* (L.).
Syst. Nat., ed. x, p. 149, 1758.
Lapland Dunlin.

382 *Calidris alpina schinzii* (C. L. Brehm).
Beitr. z. Vögelkunde, III, p. 355, 1822.
Southern Dunlin. B.

392 *Calidris testacea* (Pallas).
Vroeg's Cat. Verzam. Vogelen, etc., Adumbratiuncula, p. 6, 1764.
Saru-hama-shigi. (Curlew Sandpiper.) (R.v.)

383 *Calidris testacea* (Pallas).
Vroeg's Cat. Verzam. Vogelen, etc., Adumbratiuncula, p. 6, 1764.
Curlew-Sandpiper.

210 Tryngites Cabanis (1856)

393 *Tryngites subruficollis* (Vieillot).
Nouv. Dict. d'Hist. Nat., nouv. éd., XXXIV, p. 465, 1819.
Komon-shigi. (Buff-breasted Sandpiper.) (Exc.v.)

384 *Tryngites subruficollis* (Vieillot).
Nouv. Dict. d'Hist. Nat., nouv. éd., XXXIV, p. 465, 1819.
Buff-breasted Sandpiper. (Exc.v.)

211 Ereunetes Illiger (1811)

385 *Ereunetes pusillus pusillus* (L.).
Syst. Nat., ed. XII, p. 252, 1766.
Semi-palmated Sandpiper. (Exc.v.)

212 Philomachus Anonymous (1804)

394 *Philomachus pugnax* (L.).
Syst. Nat., ed. x, p. 148, 1758.
Erimaki-shigi. (Ruff.) (R.v.)

386 *Philomachus pugnax* (L.).
Syst. Nat., ed. x, p. 148, 1758.
Ruff. B.

213 Limicola Koch (1816)

395 *Limicola falcinellus sibirica* Dresser.
Proc. Zool. Soc. London, 1876, p. 674.
Kiriai. (Eastern Broad-billed Sandpiper.)

387 *Limicola falcinellus falcinellus* (Pontoppidan).
Danske Atlas, I, p. 623, 1763.
Broad-billed Sandpiper. (Exc.v.)

214 Limnodromus Wied (1833)

396 *Limnodromus griseus griseus* (Gmelin).
Syst. Nat., I, II, p. 658, 1789.
O-hashi-shigi. (Red-breasted Sandpiper.) (Exc.v.)

388 *Limnodromus griseus griseus* (Gmelin).
Syst. Nat., I, II, p. 658, 1789.
Red-breasted Sandpiper. (Exc.v.)

215 Eurynorhynchus Nilsson (1821)

397 *Eurynorhynchus pygmeus* (L.).
Syst. Nat., ed. x, p. 140, 1758.
Hera-shigi. (Spoon-billed Sandpiper.)

216 Terekia Bonaparte (1838)

398 *Terekia cinerea* (Güldenstädt).
Nov. Comm. Acad. Petrop., XIX, p. 473, pl. 19, 1774 or 1775.
Sorihashi-shigi. (Terek Sandpiper.)

389 *Terekia cinerea* (Güldenstädt).
Nov. Comm. Acad. Petrop., XIX, p. 473, pl. 19, 1774 or 1775.
Terek Sandpiper. (Exc.v.)

217 **Tringa** Linnæus (1758)

399 *Tringa erythropus* (Pallas).
Vroeg's Cat. Coll., Adumbratiuncula, p. 6, 1764.
Tsuru-shigi. (Spotted Redshank.)

400 *Tringa totanus eurhinus* (Oberholser).
Proc. U.S. Nat. Mus., XXII, p. 207, 1900.
Aka-ashi-shigi. (Eastern Redshank.)

401 *Tringa stagnatilis* (Bechstein).
Orn. Taschenb., II, p. 292, 1803.
Ko-awoashi-shigi. (Marsh-Sand-
piper.) (R.v.)

402 *Tringa nebularia* (Gunnerus).
Leem, Beskr. Finm. Lapp., p. 251, 1767.
Awoashi-shigi. (Greenshank.)

403 *Tringa guttifer* (Nordmann).
Ermans Reise, Naturh. Atl., p. 17, 1835.
Karafuto-awoashi-shigi. (Nord-
mann's Greenshank.) (R.v.)

404 *Tringa ochropus* L.
Syst. Nat., ed. X, p. 149, 1758.
Kusa-shigi. (Green Sandpiper.)

405 *Tringa glareola* L.
Syst. Nat., ed. X, p. 149, 1758.
Takabu-shigi. (Wood-Sandpiper.)

406 *Tringa brevipes* (Vieillot).
Nouv. Dict. d'Hist. Nat., VI, p. 410, 1816.
Kiashi-shigi. (Grey-rumped Sand-
piper.)

407 *Tringa incana* (Gmelin).
Syst. Nat., I, II, p. 658, 1789.
Meriken-kiashi-shigi. (American
Wandering Sandpiper.)

390 *Tringa erythropus* (Pallas).
Vroeg's Cat. Coll., Adumbratiuncula, p. 6, 1764.
Spotted Redshank.

391 *Tringa totanus totanus* (L.).
Syst. Nat., ed. X, p. 145, 1758.
Common Redshank. B.

392 *Tringa totanus robusta* (Schiöler).
Dansk Orn. Foren. Tidskr., XIII, p. 211, 1919.
Iceland Redshank.

393 *Tringa melanoleuca* (Gmelin).
Syst. Nat., I, II, p. 659, 1789.
Greater Yellowshank. (Exc.v.)

394 *Tringa flavipes* (Gmelin).
Syst. Nat., I, II, p. 659, 1789.
Yellowshank. (Exc.v.)

395 *Tringa stagnatilis* (Bechstein).
Orn. Taschenb., II, p. 292, 1803.
Marsh-Sandpiper. (Exc.v.)

396 *Tringa nebularia* (Gunnerus).
Leem, Beskr. Finm. Lapp., p. 251, 1767.
Greenshank. B.

397 *Tringa ochropus* L.
Syst. Nat., ed. X, p. 149, 1758.
Green Sandpiper.

398 *Tringa solitaria solitaria* Wilson.
Amer. Orn., VII, p. 53, pl. 58, fig. 3, 1813.
Solitary Sandpiper. (Exc.v.)

399 *Tringa glareola* L.
Syst. Nat., ed. X, p. 149, 1758.
Wood-Sandpiper.

400 *Tringa brevipes* (Vieillot).
Nouv. Dict. d'Hist. Nat., VI, p. 410, 1816.
Grey-rumped Sandpiper. (Exc.v.)

408 *Tringa hypoleucos* L.
Syst. Nat., ed. X, p. 149, 1758.
Iso-shigi. (Common Sandpiper.) B.

401 *Tringa hypoleucos* L.
Syst. Nat., ed. X, p. 149, 1758.
Common Sandpiper. B.

402 *Tringa macularia* L.
Syst. Nat., ed. XII, p. 249, 1766.
Spotted Sandpiper. (Exc.v.)

218 Phalaropus Brisson (1760)

409 *Phalaropus fulicarius* (L.).
Syst. Nat., ed. X, p. 148, 1758.
Haiiro-hireashi-shigi. (Grey Phalarope.)

403 *Phalaropus fulicarius* (L.).
Syst. Nat., ed. X, p. 148, 1758.
Grey Phalarope.

410 *Phalaropus lobatus* (L.).
Syst. Nat., ed. X, pp. 148, 824, 1758.
Akayeri-hireashi-shigi. (Red-necked Phalarope.) B.

404 *Phalaropus lobatus* (L.).
Syst. Nat., ed. X, pp. 148, 824, 1758.
Red-necked Phalarope. B.

219 Himantopus Brisson (1760)

405 *Himantopus himantopus himantopus* (L.).
Syst. Nat., ed. X, p. 151, 1758.
Black-winged Stilt. (R.v.)

220 Recurvirostra Linnæus (1758)

406 *Recurvirostra avosetta* L.
Syst. Nat., ed. X, p. 151, 1758.
Avocet.

221 Limosa Brisson (1760)

411 *Limosa limosa melanuroides* Gould.
Proc. Zool. Soc. London, 1846, p. 84.
Oguro-shigi. (Eastern Black-tailed Godwit.)

407 *Limosa limosa limosa* (L.).
Syst. Nat., ed. X, p. 147, 1758.
Black-tailed Godwit.

412 *Limosa lapponica baueri* Naumann.
Naturg. Vög. Deutschl., VIII, p. 429, 1836.
O-sorihashi-shigi. (Eastern Bar-tailed Godwit.)

408 *Limosa lapponica lapponica* (L.).
Syst. Nat., ed. X, p. 147, 1758.
Bar-tailed Godwit.

222 Numenius Brisson (1760)

413 *Numenius arquata lineatus* Cuvier.
Règne Animal, nouv. éd., I, p. 521, 1829.
Dai-shaku-shigi. (Indian Curlew.)

409 *Numenius arquata arquata* (L.).
Syst. Nat., ed. X, p. 145, 1758.
Common Curlew. B.

414 *Numenius tahitiensis* (Gmelin).
Syst. Nat., I, II, p. 656, 1789.
Harimomo-chu-shaku. (Bristly-thighed Curlew.) (Exc.v.)

415 *Numenius cyanopus* Vieillot.
Nouv. Dict. d'Hist. Nat., nouv. éd., VIII, p. 306, 1817.
Horoku-shigi. (Australian Curlew.)

416 *Numenius phæopus variegatus* (Scopoli).
Del. Flor. et Fauna Insubr., fasc. II, p. 92, 1786.
Chu-shaku-shigi. (Eastern Whimbrel.)

410 *Numenius phæopus phæopus* (L.).
Syst. Nat., ed. X, p. 146, 1758.
Whimbrel. B.

411 *Numenius borealis* (Forster).
Philos. Trans., LXII, p. 431, 1772.
Eskimo Curlew. (Exc.v.)

412 *Numenius tenuirostris* Vieillot.
Nouv. Dict. d'Hist. Nat., nouv. éd., VIII, p. 302, 1817.
Slender-billed Curlew. (Exc.v.)

417 *Numenius minutus* Gould.
Proc. Zool. Soc. London, 1840, p. 176, 1841.
Ko-shaku-shigi. (Little Whimbrel.)

223   Capella Frenzel (1801)

418 *Capella gallinago raddei* Buturlin.
Kuliki Rossieskoi Imperie-Premiyu-k-Journal, in Psoveia i Ruzheinaia Okhota, 1912, p. 54.
Ta-shigi. (Eastern Common Snipe.) B.

413 *Capella gallinago gallinago* (L.).
Syst. Nat., ed. X, p. 147, 1758.
Common Snipe. B.

414 *Capella gallinago faeroeensis* (C. L. Brehm).
Handb. Naturg. Vög. Deutschl., p. 617, 1831.
Færoe Snipe.

415 *Capella media* (Latham).
Gen. Syn. Suppl., I, p. 292, 1787.
Great Snipe.

419 *Capella hardwickii hardwickii* (Gray).
Zool. Misc., p. 16, 1831.
O-jishigi. (Latham's Snipe.) B.

420 *Capella megala* (Swinhoe).
Ibis, 1861, p. 343.
Chu-jishigi. (Swinhoe's Snipe.) B.

421 *Capella stenura* (Bonaparte).
"Kuhl" Ann. Stor. Nat. Bologna, IV, p. 335, 1830.
Hario-shigi. (Pin-tailed Snipe.)

422 *Capella solitaria* (Hodgson).
Gleanings in Science, III, p. 238, 1831.
Awo-shigi. (Solitary Snipe.) B.

224 **Lymnocryptes** Kaup (1829)

423 *Lymnocryptes minimus* (Brünnich).
Orn. Bor., p. 49, 1764.
    Ko-shigi. (Jack Snipe.)

416 *Lymnocryptes minimus* (Brünnich).
Orn. Bor., p. 49, 1764.
    Jack Snipe.

225 **Rhynchæa** Cuvier (1816)

424 *Rhynchæa benghalensis benghalensis*
(L.).
Syst. Nat., ed. x, p. 153, 1758.
    Tama-shigi. (Painted Snipe.) B.

226 **Scolopax** Linnæus (1758)

425 *Scolopax rusticola rusticola* L.
Syst. Nat., ed. x, p. 146, 1758.
    Yamashigi. (Woodcock.) B.

417 *Scolopax rusticola rusticola* L.
Syst. Nat., ed. x, p. 146, 1758.
    Woodcock. B.

Order LARI

Family LARIDÆ

227 **Chlidonias** Rafinesque (1822)

418 *Chlidonias niger niger* (L.).
Syst. Nat., ed. x, p. 137, 1758.
    Black Tern.

426 *Chlidonias leucopterus* (Temminck).
Man. d'Orn., p. 483, 1815.
    Hajiro-kurohara-ajisashi. (White-winged Black Tern.)

419 *Chlidonias leucopterus* (Temminck).
Man. d'Orn., p. 483, 1815.
    White-winged Black Tern. (R.v.)

420 *Chlidonias leucopareius leucopareius*
Temminck.
"Natterer" Temminck, Man. d'Orn., ed. II, II,
p. 746, 1820.
    Whiskered Tern. (Exc.v.)

228 **Gelochelidon** Brehm (1831)

421 *Gelochelidon nilotica nilotica* (Gmelin).
Syst. Nat., I, II, p. 606, 1789.
    Gull-billed Tern. (R.v.)

229 **Hydroprogne** Kaup (1829)

422 *Hydroprogne caspia* (Pallas).
Novi Comm. Acad. Petr., xiv, i, p. 582, pl. xxii,
1770.
    Caspian Tern. (R.v.)

230 Sterna Linnæus (1758)

423 *Sterna hirundo hirundo* L.
Syst. Nat., ed. x, p. 137, 1758.
Common Tern. B.

424 *Sterna macrura* Naumann.
Isis, 1819, p. 1847.
Arctic Tern. B.

425 *Sterna dougallii dougallii* Montagu.
Orn. Dict. Suppl., text and plate, 1813.
Roseate Tern. B.

427 *Sterna longipennis* Nordmann.
Ermans Verz. Thieren u. Pflanzen, p. 17, 1835.
Ajisashi. (Nordmann's Tern.) B.

428 *Sterna aleutica* Baird.
Trans. Chicago Acad. Sci., I, p. 321, pl. 31, fig. 1, 1869.
Koshijiro-ajisashi. (Aleutian Tern.)

429 *Sterna sumatrana* Raffles.
Trans. Linn. Soc. London, XIII, p. 329, 1821.
Eriguro-ajisashi. (Black-naped Tern.) (R.v.)

430 *Sterna albifrons sinensis* Gmelin.
Syst. Nat., I, II, p. 608, 1789.
Ko-ajisashi. (Asiatic Little Tern.) B.

426 *Sterna albifrons albifrons* Pallas.
In Vroeg's Cat. Verzam. Vogelen, etc., Adumbratiuncula, p. 6, 1764.
Little Tern. B.

427 *Sterna sandvicensis sandvicensis* Latham.
Gen. Syn., Suppl., I, p. 296, 1787.
Sandwich Tern. B.

431 *Sterna fuscata fuscata* (L.).
Syst. Nat., ed. XII, p. 228, 1766.
Seguro-ajisashi. (Sooty Tern.) B.

428 *Sterna fuscata fuscata* L.
Syst. Nat., ed. XII, p. 228, 1766.
Sooty Tern. (Exc.v.)

231 Melanosterna Blyth (1846)

432 *Melanosterna anaethetus anaethetus* (Scopoli).
Del. Faunae et Florae Insubr., II, p. 92, 1786.
Mamijiro-ajisashi. (Lesser Sooty Tern.) B.

232 Anous Stephens (1826)

433 *Anous stolidus pileatus* (Scopoli).
Del. Faunae et Florae Insubr., II, p. 92, 1786.
Kuro-ajisashi. (Philippine Noddy.)

233 Gygis Wagler (1832)

434 *Gygis alba kittlitzi* Hartert.
Katal. Vogelsamml. Senckenb., p. 237, 1891.
Shiro-ajisashi. (White Tern.)

234 Xema Leach (1819)

435 *Xema sabini* (Sabine).
Trans. Linn. Soc. London, XII, p. 522, pl. 29, 1818.
Kubiwa-kamome. (Sabine's Gull.)
(Exc.v.)

429 *Xema sabini* (Sabine).
Trans. Linn. Soc. London, XII, p. 522, pl. 29, 1818.
Sabine's Gull. (R.v.)

235 Rhodostethia Macgillivray (1842)

430 *Rhodostethia rosea* (Macgillivray).
Mem. Wernerian Soc., V, p. 249, 1824.
Wedge-tailed Gull. (Exc.v.)

236 Larus Linnæus (1758)

436 *Larus canus major* Middendorff.
Sibirische Reise, Zool. II, 2, p. 243, pl. xxiv, fig. 4, 1853.
Kamome. (Eastern Common Gull.)
B.

431 *Larus canus canus* L.
Syst. Nat., ed. X, p. 136, 1758.
Common Gull. B.

437 *Larus canus brachyrhynchus* Richardson.
Richardson and Swainson's Fauna Bor.-Am., II, p. 422, 1832.
Ko-kamome. (Short-billed Gull.)
(R.v.)

438 *Larus argentatus vegae* Palmén.
Vega-Exped. Vetensk. Arb., V, p. 370, 1887.
Seguro-kamome. (Palmen's Herring Gull.) B.

432 *Larus argentatus argentatus* Pontoppidan.
Danske Atlas, I, p. 622, 1763.
Herring-Gull. B.

433 *Larus argentatus cachinnans* Pallas.
Zoogr. Rosso-Asiat., II, p. 318, 1827.
Yellow-legged Herring-Gull.
(Exc.v.)

439 *Larus schistisagus* Stejneger.
Auk, 1884, p. 231.
O-seguro-kamome. (Slaty-backed Gull.) B.

434 *Larus marinus* L.
Syst. Nat., ed. X, p. 136, 1758.
Great Black-backed Gull. B.

435 *Larus fuscus fuscus* L.
Syst. Nat., ed. X, p. 136, 1758.
Scandinavian Lesser Black-backed Gull.

H.

436 *Larus fuscus affinis* Reinhardt.
Vidensk. Meddl. Kjöbenhavn for 1853, p. 78, 1854.
British Lesser Black-backed Gull. B.

440 *Larus crassirostris* Vieillot.
Nouv. Dict. d'Hist. Nat., XXI, p. 508, 1818.
Umineko. (Black-tailed Gull.) B.

441 *Larus glaucescens* Naumann.
Vög. Deutschl., X, p. 351, 1840.
Washi-kamome. (Glaucous-winged Gull.)

442 *Larus hyperboreus* Gunnerus.
Leem's Beskr. Finm. Lapp., pp. 226, 283, 1767.
Shiro-kamome. (Glaucous Gull.) B.

437 *Larus hyperboreus* Gunnerus.
Leem's Beskr. Finm. Lapp., pp. 226, 283, 1767.
Glaucous Gull.

443 *Larus glaucoides* Meyer.
Zusätze u. Ber. zu Meyer and Wolf's Taschenb.
d. deutsch. Vögelk., p. 197, 1822.
Hajiro-kamome. (Iceland Gull.)

438 *Larus glaucoides* Meyer.
Zusätze u. Ber. zu Meyer and Wolf's Taschenb.
d. deutsch. Vögelk., p. 197, 1822.
Iceland Gull.

439 *Larus ichthyaëtus* Pallas.
Reise d. versch. Prov. d. Russ. Reichs, II, p. 713, 1773.
Great Black-headed Gull. (Exc.v.)

440 *Larus melanocephalus* Temminck.
Natterer, Isis, 1818, p. 816.
Mediterranean Black-headed Gull. (Exc.v.)

441 *Larus philadelphia* (Ord.).
In Guthrie's Geogr., 2nd Amer. ed., p. 319, 1815.
Bonaparte's Gull. (Exc.v.)

444 *Larus ridibundus ridibundus* L.
Syst. Nat., ed. XII, p. 225, 1766.
Yuri-kamome. (Black-headed Gull.) B.

442 *Larus ridibundus ridibundus* L.
Syst. Nat., ed. XII, p. 225, 1766.
Black-headed Gull. B.

445 *Larus ridibundus sibiricus* Buturlin.
Mess. Orn., II, p. 66, 1911.
O-yuri-kamome. (Siberian Black-headed Gull.)

446 *Larus saundersi* (Swinhoe).
Proc. Zool. Soc. London, 1871, pp. 273, 421, pl. xxii.
Zuguro-kamome. (Saunder's Gull.)

443 *Larus minutus* Pallas.
Reise d. versch. Prov. d. Russ. Reichs, III, p. 702, 1776.
Little Gull.

237 Pagophila Kaup (1829)

444 *Pagophila eburnea* (Phipps).
Voy. N. Pole, App., p. 187, 1774.
Ivory-Gull. (R.v.)

238 Rissa Stephens (1826)

447 *Rissa tridactyla pollicaris* Stejneger.
Baird, Brewer and Ridgw., Water B.N. Amer.,
II, p. 202, 1884.
Mitsuyubi-kamome. (Pacific Kitti-
wake Gull.)

445 *Rissa tridactyla tridactyla* (L.).
Syst. Nat., ed. x, p. 136, 1758.
Kittiwake Gull. B.

239 Catharacta Brünnich (1764)

446 *Catharacta skua skua* Brünnich.
Orn. Bor., p. 33, 1764.
Great Skua. B.

448 *Catharacta matsudairae* Takatsukasa.
"Tori," Vol. III, pp. 105, 107, 1922.
O-tozoku-kamome. (Japanese Great
Skua.)

240 Stercorarius Brisson (1760)

449 *Stercorarius longicaudus* Vieillot.
Nouv. Dict. d'Hist. Nat., nouv. éd., XXXII, p.
157, 1819.
Shirohara-tozoku-kamome. (Long-
tailed Skua.)

447 *Stercorarius longicaudus* Vieillot.
Nouv. Dict. d'Hist. Nat., nouv. éd., XXXII, p.
157, 1819.
Long-tailed Skua.

450 *Stercorarius parasiticus* (L.).
Syst. Nat., ed. x, p. 136, 1758.
Kuro-tozoku-kamome. (Arctic
Skua.)

448 *Stercorarius parasiticus* (L.).
Syst. Nat., ed. x, p. 136, 1758.
Arctic Skua. B.

451 *Stercorarius pomarinus* (Temminck).
Man. d'Orn., p. 514, 1815.
Tozoku-kamome. (Pomatorhine
Skua.)

449 *Stercorarius pomarinus* (Temminck).
Man. d'Orn., p. 514, 1815.
Pomatorhine Skua.

Order ALCÆ

Family ALCIDÆ

241 Alca Linnæus (1758)

450 *Alca torda* L.
Syst. Nat., ed. x, p. 130, 1758.
Razorbill. B.

451 *Alca impennis* L.
Syst. Nat., ed. x, p. 130, 1758.
Great Auk. (Extinct.)

242 **Alle** Link (1806)

452 *Alle alle* (L.).
Syst. Nat., ed. x, p. 131, 1758.
Little Auk.

243 **Uria** Brisson (1760)

452 *Uria aalge californica* (Bryant).
Proc. Boston Soc. N. Hist., VIII, p. 142, 1861.
Umi-garasu. (Californian Guillemot.) B.

453 *Uria aalge aalge* (Pontoppidan).
Danske Atlas, I, pl. xxvi, and p. 621, 1763.
Northern Guillemot.

454 *Uria aalge albionis* Witherby.
Brit. Birds (Mag.), XVI, p. 324, 1923.
Southern Guillemot. B.

453 *Uria lomvia arra* (Pallas).
Zoogr. Rosso-Asiat., II, p. 347, 1827.
Hashibuto-umigarasu. (Pallas's Guillemot.) B.

455 *Uria lomvia lomvia* (L.).
Syst. Nat., ed. x, p. 130, 1758.
Brünnich's Guillemot. (R.v.)

456 *Uria grylle grylle* (L.).
Syst. Nat., ed. x, p. 130, 1758.
Black Guillemot. B.

244 **Cepphus** Pallas (1769)

454 *Cepphus carbo* Pallas.
Zoogr. Rosso-Asiat., II, p. 350, 1827.
Keimafuri. (Sooty Guillemot.) B.

455 *Cepphus columba snowi* Stejneger.
Auk, 1897, p. 201.
Chishima-umibato.   (Stejneger's Pigeon Guillemot.)

456 *Cepphus columba columba* Pallas.
Zoogr. Rosso-Asiat., II, p. 348, 1827.
Umibato. (Pigeon Guillemot.)

245 **Brachyramphus** Brandt (1837)

457 *Brachyramphus marmoratus perdix* (Pallas).
Zoogr. Rosso-Asiat., II, p. 351, pl. lxxx, 1827.
Madara-umisuzume. (Partridge Auk.)

458 *Brachyramphus brevirostris* (Vigors).
Zool. Journ., IV, p. 357, 1828.
Kobashi-umisuzume.   (Short-billed Auk.)

246 **Ptychorhamphus** Brandt (1837)

459 *Ptychorhamphus aleuticus* (Pallas).
Zoogr. Rosso-Asiat., II, p. 370, 1827.
America-umisuzume. (Cassin's Auk.)

247 **Synthliboramphus** Brandt (1837)

460 *Synthliboramphus antiquus* (Gmelin).
　Syst. Nat., I, ii, p. 554, 1789.
　　Umisuzume. (Ancient Auk.) B.

461 *Synthliboramphus wumizusume* (Tem-
　minck).
　Pl. Color., 579, 1835.
　　Kammuri-umisuzume. (Japanese
　　Auk.) B.

248 **Aethia** Merrem (1788)

462 *Aethia pusilla* (Pallas).
　Zoogr. Rosso-Asiat., ii, p. 373, 1827.
　　Ko-umisuzume. (Least Auk.)

463 *Aethia pygmaea* (Gmelin).
　Syst. Nat., I, ii, p. 555, 1789.
　　Shirahige-umisuzume. (Whiskered
　　Auk.)

464 *Aethia cristatella* (Pallas).
　Spicil. Zool., fasc. v, p. 18, pl. iii, and v, figs. 7–9,
　1769.
　　Etorofu-umisuzume. (Crested Auk.)
　　B.

249 **Phaleris** Temminck (1820)

465 *Phaleris psittacula* (Pallas).
　Spicil. Zool., fasc. v, p. 13, pl. ii, and v, figs. 4–6,
　1769.
　Umi-omu. (Parroquet Auk.)

250 **Cerorhinca** Bonaparte (1828)

466 *Cerorhinca monocerata* (Pallas).
　Zoogr. Rosso-Asiat., ii, p. 362, 1827.
　　Uto. (Hornbilled Puffin.) B.

251 **Lunda** Pallas (1827)

467 *Lunda cirrhata* (Pallas).
　Spicil. Zool., fasc. v, p. 7, pl. i, and v, figs. 1–3,
　1769.
　　Etopirika. (Tufted Puffin.) B.

252 **Fratercula** Brisson (1760)

457 *Fratercula arctica grabæ* (C. L. Brehm).
　Handb. Naturg. Vög. Deutschl., p. 999, 1831.
　　Southern Puffin. B.

468 *Fratercula corniculata* (Naumann).
　Isis, 1821, p. 782, pl. 7, figs. 3, 4.
　　Tsunomedori. (Horned Puffin.)

Order TUBINARES

Family PROCELLARIIDÆ

253 **Hydrobates** Boie (1822)

458 *Hydrobates pelagica* (L.).
Syst. Nat., ed. X, p. 131, 1758.
Storm-Petrel. B.

254 **Oceanodroma** Reichenbach (1852)

469 *Oceanodroma leucorrhoa leucorrhoa*
(Vieillot).
Nouv. Dict. d'Hist. Nat., nouv. éd., XXV, p. 422,
1817.
Koshijiro-umitsubame. (Leach's
Fork-tailed Petrel.) B.

470 *Oceanodroma castro* (Harcourt).
Sketch of Madeira, p. 123, 1851.
Kuro-koshijiro-umitsubame.
(Madeiran Fork-tailed Petrel.)
(Exc.v.)

471 *Oceanodroma markhami owstoni*
(Mathews and Iredale).
Ibis, 1915, p. 581.
Owston-umitsubame. (Owston's
Fork-tailed Petrel.)

472 *Oceanodroma melania matsudairae*
Kuroda.
Ibis, 1922, p. 311.
Kuro-umitsubame. (Kuroda's Fork-
tailed Petrel.)

473 *Oceanodroma monorhis monorhis* (Swin-
hoe).
Ibis, 1867, p. 386, 1869, p. 348.
Hime-kuro-umitsubame. (Swinhoe's
Fork-tailed Petrel.) B.

474 *Oceanodroma furcata* (Gmelin).
Syst. Nat., I, II, p. 561, 1789.
Haiiro-umitsubame. (Grey Fork-
tailed Petrel.)

459 *Oceanodroma leucorrhoa leucorrhoa*
(Vieillot).
Nouv. Dict. d'Hist. Nat., nouv. éd., XXV, p. 422,
1817.
Leach's Fork-tailed Petrel. B.

460 *Oceanodroma castro* (Harcourt).
Sketch of Madeira, p. 123, 1851.
Madeiran Fork-tailed Petrel. (Exc.v.)

255 **Oceanites** Keyserling and Blasius (1840)

461 *Oceanites oceanicus* (Kuhl).
Beiträge z. Zool., p. 136, pl. 10, fig. 1, 1820.
Wilson's Petrel. (Exc.v.)

256 Pelagodroma Reichenbach (1852)

462 *Pelagodroma marina hypoleuca* (Webb, Berthelot and Moquin-Tandon).
Hist. Nat. Iles Canar., Zool., Orn., p. 45, 1841.
Frigate-Petrel.

257 Puffinus Brisson (1760)

463 *Puffinus puffinus puffinus* (Brünnich).
Orn. Bor., p. 29, 1764.
Manx Shearwater. B.

464 *Puffinus puffinus mauretanicus* Lowe.
Bull. B. O. C., XLI, p. 140, 1921.
Western Mediterranean Shearwater.
(R.v.)

475 *Puffinus leucomelas* (Temminck).
Pl. Color., 587, livr. 99 c, 1836.
O-mizunagidori. (Streaked Shearwater.) B.

476 *Puffinus cuneatus* Salvin.
Ibis, 1888, p. 353.
Onaga-mizunagidori. (Salvin's Shearwater.) (R.v.)

477 *Puffinus tenuirostris tenuirostris* (Temminck).
Pl. Color., text to pl. 587, 1835.
Hashiboso-mizunagidori. (Slender-billed Shearwater.) B.

478 *Puffinus griseus* (Gmelin).
Syst. Nat., I, II, p. 564, 1789.
Haiiro-mizunagidori. (Sooty Shearwater.) B.

465 *Puffinus griseus* (Gmelin).
Syst. Nat., I, II, p. 564, 1789.
Sooty Shearwater.

466 *Puffinus gravis* (O'Reilly).
Greenland, adjacent seas, etc., p. 140, pl. 12, fig. 1, 1818.
Great Shearwater.

467 *Puffinus kuhlii kuhlii* (Boie).
Isis, 1835, p. 257.
Mediterranean Great Shearwater.
(Exc.v.)

468 *Puffinus kuhlii borealis* Cory.
Bull. Nuttall Orn. Club, VI, p. 84, 1881.
North Atlantic Great Shearwater.
(Exc.v.)

469 *Puffinus assimilis baroli* Bonaparte.
Consp. Gen. Av., II, p. 204, 1857.
Madeiran Little Shearwater.
(Exc.v.)

470 *Puffinus assimilis boydi* Mathews.
  B. Australia, II, p. 70, 1912.
    Cape Verde Little Shearwater.
    (Exc.v.)

479 *Puffinus carneipes* Gould.
  Ann. and Mag. Nat. Hist., XIII, p. 365, 1844.
    Akaashi-mizunagidori. (Japanese
    Pink-footed Shearwater.) (Exc.v.)

258 Pterodroma Bonaparte (1856)

471 *Pterodroma hasitata* (Kuhl).
  Beiträge z. Zool., 2 Abt., p. 142, 1820.
    Capped Petrel. (Exc.v.)

472 *Pterodroma brevipes* (Peale).
  U.S. Expl. Exp., VIII, pp. 294, 337, pl. 80, 1848.
    Collared Petrel. (Exc.v.)

473 *Pterodroma neglecta* (Schlegel).
  Mus. Pays-Bas, VI Procell., p. 10, 1863.
    Kermadec Petrel. (Exc.v.)

480 *Pterodroma hypoleuca* (Salvin).
  Ibis, 1888, p. 359.
    Shirohara-mizunagidori. (Bonin
    Island Fulmar.)

481 *Pterodroma longirostris* (Stejneger).
  Proc. U.S. Nat. Mus., XVI, p. 618, 1893.
    Hime-shirohara-mizunagidori.
    (Japanese Fulmar.)

259 Bulweria Bonaparte (1843)

482 *Bulweria bulwerii* (Jardine and Selby).
  Illustr. Orn., II, pl. 65 and text, 1828.
    Anadori. (Bulwer's Petrel.)

474 *Bulweria bulwerii* (Jardine and Selby).
  Illustr. Orn., II, pl. 65 and text, 1828.
    Bulwer's Petrel. (R.v.)

260 Fulmarus Stephens (1826)

483 *Fulmarus glacialis rodgersii* Cassin.
  Proc. Acad. Philadelphia, 1862, p. 326.
    Furuma-kamome. (Pacific Fulmar.)

475 *Fulmarus glacialis glacialis* (L.).
  Fauna Svecica, ed. II, p. 51, 1761.
    Fulmar Petrel. B.

261 Diomedea Linnæus (1758)

476 *Diomedea melanophrys* Temminck.
  (ex Boie MS.), Pl. Color., 456, 1828.
    Black-browed Albatros. (Exc.v.)

484 *Diomedea albatrus* Pallas.
  Spicil. Zool., V, p. 28, 1780.
    Ahodori. (Steller's Albatros.)

485 *Diomedea immutabilis* Rothschild.
  Bull. B. O. C., vol. I, p. xlviii, 1893.
    Ko-ahodori. (Laysan Albatros.)

486 *Diomedea nigripes* Audubon.
  Orn. Biogr., V, p. 327, 1839.
    Kuroashi-ahodori. (Black-footed
    Albatros.)

Order PYGOPODES
Family PODICIPIDÆ
262 **Podiceps** Latham (1787)

487 *Podiceps cristatus cristatus* (L.).
Syst. Nat., ed. x, p. 135, 1758.
Kammuri-kaitsuburi. (Great-crested Grebe.)

488 *Podiceps griseigena holboellii* Reinhardt.
Videnskab. Middelelser, 1853, p. 76.
Akaeri-kaitsuburi. (Eastern Red-necked Grebe.) B.

489 *Podiceps auritus* (L.).
Syst. Nat., ed. x, p. 135, 1758.
Mimi-kaitsuburi. (Slavonian Grebe.)

490 *Podiceps nigricollis nigricollis* C. L. Brehm.
Handb. Naturg. Vög. Deutschl., p. 963, 1831.
Hajiro-kaitsuburi. (Black-necked Grebe.)

491 *Podiceps ruficollis japonicus* Hartert.
Vog. Pal. Faun., II, p. 1455, 1920.
Kaitsuburi. (Japanese Little Grebe.) B.

477 *Podiceps cristatus cristatus* (L.).
Syst. Nat., ed. x, p. 135, 1758.
Great-crested Grebe. B.

478 *Podiceps griseigena griseigena* (Boddaert).
Tabl. Pl. Enl., p. 55, 1783.
Red-necked Grebe.

479 *Podiceps auritus* (L.).
Syst. Nat., ed. x, p. 135, 1758.
Slavonian Grebe. B.

480 *Podiceps nigricollis nigricollis* C. L. Brehm.
Handb. Naturg. Vög. Deutschl., p. 963, 1831
Black-necked Grebe. B.

481 *Podiceps ruficollis ruficollis* (Pallas).
Vroeg's Cat. Coll., Adumbratiuncula, p. 6, 1764.
Little Grebe. B.

Family COLYMBIDÆ
263 **Colymbus** Linnæus (1758)

492 *Colymbus arcticus viridigularis* (Dwight).
Auk, 1918, p. 198.
Ohamu. (Siberian Black-throated Diver.)

493 *Colymbus arcticus pacificus* Lawrence.
Baird's B. N. Amer., p. 889, 1860.
Shiroeri-ohamu. (Pacific Black-throated Diver.)

494 *Colymbus adamsii* Gray.
Proc. Zool. Soc. London, 1859, p. 167.
Hashijiro-abi. (White-billed Northern Diver.)

495 *Colymbus stellatus* Pontoppidan.
Danske Atlas, I, p. 621, 1763.
Abi. (Red-throated Diver.) B.

482 *Colymbus arcticus arcticus* L.
Syst. Nat., ed. x, p. 135, 1758.
Black-throated Diver. B.

483 *Colymbus immer* Brünnich.
Orn. Bor., p. 38, 1764.
Great Northern Diver.

484 *Colymbus adamsii* Gray.
Proc. Zool. Soc. London, 1859, p. 167.
White-billed Northern Diver. (Exc.v.)

485 *Colymbus stellatus* Pontoppidan.
Danske Atlas, I, p. 621, 1763.
Red-throated Diver. B.

Order COLUMBÆ
Family COLUMBIDÆ

264 Columba Linnæus (1758)

486 *Columba œnas* L.
Syst. Nat., ed. x, p. 162, 1758.
Stock-Dove. B.

487 *Columba palumbus palumbus* L.
Syst. Nat., ed. x, p. 163, 1758.
Wood-Pigeon. B.

488 *Columba livia livia* Gmelin.
Syst. Nat., I, II, p. 769, 1789.
Rock-Dove. B.

265 Janthœnas Bonaparte (1854)

496 *Janthœnas janthina janthina* (Temminck).
Pl. Color., 503, livr. 86, 1830.
Karasu-bato. (Japanese Fruit-Pigeon.) B.

266 Streptopelia Bonaparte (1854)

489 *Streptopelia turtur turtur* (L.).
Syst. Nat., ed. x, p. 164, 1758.
Turtle-Dove. B.

497 *Streptopelia decaocto decaocto* (Frivalszky).
A. M. Társaság Erkönyvei (Ungarische Akademieschriften) 1834–1836, Vol. 3, part 3, pp. 183, 184, pl. 8, 1838.
Shirako-bato or Juzukake-bato. (Indian Ring-Dove.) B.

498 *Streptopelia orientalis orientalis* (Latham).
Index Orn., II, p. 606, 1790.
Kiji-bato. (Eastern Rufous Turtle-Dove.) B.

490 *Streptopelia orientalis orientalis* (Latham).
Index Orn., II, p. 606, 1790.
Eastern Rufous Turtle-Dove. (Exc.v.)

267 Sphenurus Swainson (1837)

499 *Sphenurus sieboldii sieboldii* (Temminck).
Pl. Color., 549, livr. 93, 1835.
Awobato. (Japanese Green-Pigeon.) B.

500 *Sphenurus permagnus permagnus* (Stejneger).
Proc. U.S. Nat. Mus., IX, p. 637, 1887.
Riukiu-awo-bato. (Loo-Choo Island Green Pigeon.)

Order PTEROCLETES

268 Syrrhaptes Illiger (1811)

501 *Syrrhaptes paradoxus* (Pallas).
Reise d. versch. Prov. d. Russ. Reichs, II, p. 712, pl. F, 1773.
Sakei. (Pallas's Sand-Grouse.) (Exc.v.)

491 *Syrrhaptes paradoxus* (Pallas).
Reise d. versch. Prov. d. Russ. Reichs, II, p. 712, pl. F, 1773.
Pallas's Sand-Grouse. (R.v.)

Order HEMIPODII

Family TURNICIDÆ

269 Turnix Bonnaterre (1790)

502 *Turnix javanica blakistoni* (Swinhoe).
Proc. Zool. Soc. London, p. 401, 1871.
Mifu-uzura. (Chinese Hemipode.) B.

Order GALLI

Family TETRAONIDÆ

270 Tetrao Linnæus (1758)

492 *Tetrao urogallus urogallus* L.
Syst. Nat., ed. X, p. 159, 1758.
Capercaillie. B.

503 *Tetrao parvirostris parvirostris* Bonaparte.
Compt. Rend. Acad. Paris, XLII, p. 880, 1856.
O-raicho. (Siberian Capercaillie.) B.

271 Lyrurus Swainson (1832)

493 *Lyrurus tetrix britannicus* Witherby and Lonnberg.
Brit. B. (Mag.), VI, p. 270, 1913.
British Black-Grouse. B.

272 Lagopus Brisson (1760)

494 *Lagopus scoticus scoticus* (Latham).
Gen. Syn., Suppl., I, p. 290, 1787.
British Red Grouse. B.

495 *Lagopus scoticus hibernicus* (Kleinschmidt).
Falco, XV, p. 3, 1919.
Irish Red Grouse. B.

504 *Lagopus mutus kurilensis* Kuroda.
Bull. B.O.C. CCXC, p. 15, Oct. 1924.
Chishima-raicho. (Kuril Is. Ptarmigan.) B.

496 *Lagopus mutus millaisi* Hartert.
Brit. B. (Mag.), XVII, p. 106, 1923.
Scottish Ptarmigan. B.

505 *Lagopus mutus japonicus* Clark.
Proc. U.S. Nat. Mus., XXXII, p. 469, 1907.
Raicho. (Japanese Ptarmigan.) B.

506 *Lagopus lagopus lagopus* (L.).
Syst. Nat., ed. X, p. 159, 1758.
Karafuto-raicho. (Willow Grouse.)
B.

273 Falcipennis Elliot (1864)

507 *Falcipennis falcipennis* (Hartlaub).
Journ. f. Orn., 1855, p. 39.
Kamabane-raicho. (Siberian Spruce
Grouse.) B.

274 Tetrastes Keyserling and Blasius (1840)

508 *Tetrastes bonasia vicinitas* Riley.
Proc. Biol. Soc. Washington, XXVIII, p. 161,
1915.
Yezo-raicho. (Japanese Hazel
Grouse.) B.

509 *Tetrastes bonasia kolymensis* Buturlin.
Mess. Orn. Dez., 1916, p. 226.
Karafuto yezo-raicho. (Sakhalien
Island Hazel Grouse.) B.

Family PHASIANIDÆ

275 Phasianus Linnæus (1758)

497 *Phasianus colchicus* L.
Syst. Nat., ed. X, p. 158, 1758.
Pheasant. B.

510 *Phasianus versicolor versicolor* Vieillot.
Gal. Ois., II, p. 23, pl. 205, 1825.
Kita-kiji. (Northern Green Pheasant.)
B.

511 *Phasianus versicolor robustipes* Kuroda.
Dobutsu. Zasshi (Tokio Zool. Mag.), XXXI, pp.
299, 309, 1919.
Hokuroku-kiji. (Western Green
Pheasant.) B.

512 *Phasianus versicolor tohkaidi* Momi-
yama.
Dobutsu. Zasshi (Tokio Zool. Mag.), 1922, p.
734.
Ma-kiji. (Tokaido Green Pheasant.)
B.

513 *Phasianus versicolor affinis* Momiyama.
Dobutsu. Zasshi (Tokio Zool. Mag.), 1922, p. 165.
Ijima-kiji. (Ijima's Green Pheasant.)
B.

514 *Phasianus versicolor maedaius* Momi-yama.
Dobutsu. Zasshi (Tokio Zool. Mag.), 1922, p. 736.
Tamba-kiji. (Western Hondo Green Pheasant.) B.

515 *Phasianus versicolor nankaidi* Momi-yama.
Dobutsu. Zasshi (Tokio Zool. Mag.), 1922, p. 737.
Shikoku-kiji. (Shikoku Green Pheasant.) B.

516 *Phasianus versicolor kiusiuensis* Ku-roda.
Dobutsu. Zasshi (Tokio Zool. Mag.), XXXI, p. 309, 1919.
Kiusiu-kiji. (Kiusiu Green Pheasant.) B.

517 *Phasianus versicolor tanensis* Kuroda.
Dobutsu. Zasshi (Tokio Zool. Mag.), XXXI, p. 310, 1919.
Shima-kiji. (Southern Green Pheasant.) B.

276 Syrmaticus Wagler (1832)

518 *Syrmaticus sœmmerringii sœmmer-ringii* (Temminck).
Pl. Color., 487, 488, 1830.
Aka-yamadori. (Copper Pheasant.) B.

519 *Syrmaticus sœmmerringii scintillans* (Gould).
Ann. and Mag. Nat. Hist., (3) XVII, p. 150, 1866.
Yamadori. (Hondo Copper Pheasant.) B.

520 *Syrmaticus sœmmerringii intermedius* (Kuroda).
Dobutsu. Zasshi (Tokio Zool. Mag.), XXXI, p. 312, 1919.
Shikoku-yamadori. (Shikoku Copper Pheasant.) B.

521 *Syrmaticus sœmmerringii subrufus*
(Kuroda).
Dobutsu. Zasshi (Tokio Zool. Mag.), XXXI, p.
312, 1919.
Usuaka-yamadori. (Lesser Copper
Pheasant.) B.

522 *Syrmaticus sœmmerringii ijimae*
(Dresser).
Ibis, 1902, p. 656.
Koshijiro-yamadori. (White-rumped
Copper Pheasant.) B.

277 **Bambusicola** Gould (1862)

523 *Bambusicola thoracica thoracica*
(Temminck).
Temm. Pig. et Gall., III, pp. 375 and 723, 1815.
Ko-jukei. (Chinese Bamboo-
Pheasant.) B. (Introduced.)

278 **Perdix** Brisson (1760)

498 *Perdix perdix perdix* (L.).
Syst. Nat., ed. X, p. 160, 1758.
Common Partridge. B.

279 **Coturnix** Bonnaterre (1791)

524 *Coturnix coturnix japonica* Temminck
and Schlegel.
Siebold's Fauna Japon., Aves, p. 103, pl. 61,
1849.
Uzura. (Japanese Quail.) B.

499 *Coturnix coturnix coturnix* (L.).
Syst. Nat., ed. X, p. 161, 1758.
Quail. B.

280 **Alectoris** Kaup (1829)

500 *Alectoris rufa rufa* (L.).
Syst. Nat., ed. X, p. 169, 1758.
Red-legged Partridge. B.

INDEX

Japanese common names are in italic type, and the names of Birds which occur in both countries are in brackets.